Optical Cryptosystems

SERIES EDITOR

 **Professor Rajpal S Sirohi** Consultant Scientist

About the Editor

Rajpal S Sirohi is currently working as a faculty member in the Department of Physics, Alabama A&M University, Huntsville, Alabama (USA). Prior to this, he was a consultant scientist at the Indian Institute of Science Bangalore, and before that he was chair professor in the Department of Physics, Tezpur University, Assam. During 2000–11, he was academic administrator, being vice chancellor to a couple of universities and the director of the Indian Institute of Technology Delhi. He is the recipient of many international and national awards and the author of more than 400 papers. Dr Sirohi is involved with research concerning optical metrology, optical instrumentation, holography, and speckle phenomenon.

About the series

Optics, photonics and optoelectronics are enabling technologies in many branches of science, engineering, medicine and agriculture. These technologies have reshaped our outlook, our way of interaction with each other and brought people closer. They help us to understand many phenomena better and provide a deeper insight in the functioning of nature. Further, these technologies themselves are evolving at a rapid rate. Their applications encompass very large spatial scales from nanometers to astronomical and a very large temporal range from picoseconds to billions of years. The series on the advances on optics, photonics and optoelectronics aims at covering topics that are of interest to both academia and industry. Some of the topics that the books in the series will cover include bio-photonics and medical imaging, devices, electromagnetics, fiber optics, information storage, instrumentation, light sources, CCD and CMOS imagers, metamaterials, optical metrology, optical networks, photovoltaics, freeform optics and its evaluation, singular optics, cryptography and sensors.

About IOP ebooks

The authors are encouraged to take advantage of the features made possible by electronic publication to enhance the reader experience through the use of colour, animation and video, and incorporating supplementary files in their work.

Do you have an idea of a book you'd like to explore?

For further information and details of submitting book proposals see iopscience.org/books or contact Ashley Gasque on Ashley.gasque@iop.org.

Optical Cryptosystems

Naveen K Nishchal

Department of Physics, Indian Institute of Technology Patna, Patna, Bihar, India

IOP Publishing, Bristol, UK

ISBN   978-0-7503-2220-1 (ebook)
ISBN   978-0-7503-2218-8 (print)
ISBN   978-0-7503-2221-8 (myPrint)
ISBN   978-0-7503-2219-5 (mobi)

DOI   10.1088/978-0-7503-2220-1

Version: 20191201

IOP ebooks

British Library Cataloguing-in-Publication Data: A catalogue record for this book is available from the British Library.

Published by IOP Publishing, wholly owned by The Institute of Physics, London

IOP Publishing, Temple Circus, Temple Way, Bristol, BS1 6HG, UK

US Office: IOP Publishing, Inc., 190 North Independence Mall West, Suite 601, Philadelphia, PA 19106, USA

असतो मा सदगमय ॥ तमसो मा ज्योतिर्गमय ॥ मृत्योर्मामृतम् गमय ॥

- बृहदारण्यक उपनिषद् 1.3.27

'Asato ma sadgamaya, Tamaso ma jyotirgamaya, Mrityorma amritamgamaya'

Oh Almighty! Lead us from the unreal (falsity) to the real (truth),
From darkness to light!
From death to immortality!
–Brihdaranyaka Upanisada 1:3:27 - India

अप्प दीपो भव।

"Appa Deepo Bhavah"
Be a Light unto Yourself.

Gautama Buddha

Dedicated to the memory of my parents
Shrimati Kamini Devi and Shri Balram Prasad Singh

Contents

Preface

In the digital era of contemporary society, information in any form, such as a message, text, data, image, audio, or video, can be treated as wealth. Therefore, securing information is as important as protecting property. In the history of the human race, the significance of security in one form or the other can easily be traced. Though cryptographic techniques have been in use for protecting information for thousands of years, the systematic study of cryptology as a science started around one hundred years ago. Julius Caesar (around 100 BC) was known to use a form of encryption to convey secret messages to his Army Generals. In modern times, digital techniques of information security are already in use wherein there exists scope for further improvements in terms of security level and computation cost.

Owing to the unique features of light, such as parallel processing, high speed, and several degrees of freedom, it is envisaged that information can be highly secured and communicated to the intended recipients or authentic users employing optical technologies. It can be foreseen that with the multifaceted uses of advanced technologies, such as *Artificial Intelligence*, *Big Data*, *Cloud Computing*, and *Internet-of-Things*, security will always remain an important challenge. Technologies provide several opportunities, but, at the same time, they also pose threats to information theft or misuse. Searching for a cyber expert or the attackers who attacked the digital algorithm would be very hard, because they can exist in large numbers anywhere in the world. On the other hand, finding out an attacker in the optics domain would be relatively easier. The security can be in terms of storage, in dissemination of the message, communication/transmission over conventional channels, protection of copyright/ownership, and steganography. Therefore, developments of newer alternative technologies are required to meet the challenges in the domains of scientific investigation.

This book intends to provide a collection of optical technologies for secure storage, secure communication, and the protection of copyright in terms of watermarking. Most of the optical techniques reported in literature can be traced around a double random phase encoding algorithm. Furthermore, many variants of this scheme have been proposed and demonstrated with improvements and different levels of complexity. This book aims to provide help to researchers in the field to get first-hand information of its progress.

This book starts with a general discussion on digital algorithms already in use in chapter 1 with more emphasis on the principles of optical techniques for image/data security in chapter 2. The growth of literature on optical technologies has been exponential with the publication of the first report in 1995. A bar chart has been provided that shows the growth of the literature. Use of fully-phased data provides additional security and robustness against noise, therefore such techniques have been dealt with in chapter 3. There is another aspect associated with security that is called authentication, in which the retrieval of original information is not intended. This can be solved with the use of an optical correlator, called a joint transform correlator, which is discussed in chapter 4. Optical techniques of watermarking and hiding are discussed in chapter 5. Polarization is one of the important properties of light, which is suited to developing a practical system because in this case the parameter that is dealt with is intensity, not the phase. Therefore, storage and transmission of intensity data is easier than phase-only information. This has been detailed in chapter 6.

Digital holography helps record 3D data and recording with digital sensors offers advantages in image/data security. The digital holograms can be stored in a personal computer and transmitted anywhere in the world and can be numerically reconstructed at any point of time. This has been discussed in chapter 7. Processing and security of multispectral data is very important in many applications, particularly in defence, remote sensing, and surveillance. This has been discussed in chapter 8. Chaos has always been very attractive in cryptographic studies in key design. Chapter 9 has been devoted to this topic, which has the ability to combine with other optical technologies. Phase retrieval techniques are important in regenerating object-dependent phase keys used for securing data. There are different algorithms reported in literature, which find use in image security. This has been dealt with in chapter 10.

No cryptographic technique can be considered very strong and useful unless cryptanalysis is carried out. There are several types of attacks reported in literature, which have been stated in terms of optical technologies in chapter 11. The optical technologies differ with digital counterparts, whereby in optical schemes either physical keys are used or keys are designed considering physical parameters as compared to digital keys used in electronic systems. There are various types of keys implemented in optical methods, which are discussed in chapter 12.

In all the chapters, the basic principles have been explained with examples. In some of the chapters, numerical simulation results have been provided for better understanding of the subject. Considering the requirement on some of the relevant topics, MATLAB codes have been provided. At the end of each chapter, a list of relevant literature has been provided.

The book is open to comments, criticisms, and suggestions from the readers in improving the quality of the book for future editions.

Acknowledgement

First of all, I would like to thank my mentors at the Indian Institute of Technology (IIT) Delhi during my doctoral studies. I especially thank my PhD supervisors, Professor Kehar Singh and Professor Joby Joseph, for introducing me to the exciting world of optical information processing research. Over the years they have been much more than just a thesis advisor to me. I am sure that they will be happy to see this book out in print. I must thank Dr G Unnikrishnan from IRDE Dehradun with whom I have had numerous discussions. I wish to extend my sincere thanks to Dr A K Gupta, Ex-Director IRDE Dehradun and Professor R S Sirohi, Ex-Director, IIT Delhi, for their encouragement.

My colleagues at IIT Patna deserve due acknowledgments for their encouragement. I acknowledge my gratitude to IIT Patna for providing facilities and a congenial environment. I must thank my colleagues from abroad, Professor Bahram Javidi, University of Connecticut; Professor John T Sheridan, University College Dublin; Professor Thomas J Naughton, Maynooth University; Professor Ayman AlFalou, ISEN/Yncrea; Professor Christian Brosseau, Universite Britagne Occidentale; Professor Cornelia Denz, Universitat Muenster; Professor Yan Zhang, Capital Normal University; Professor Takanori Nomura, Wakayama University; Professor Osamu Matoba, Kobe University; Professor Guohai Situ, Shanghai Institute of Optics and Fine Mechanics; and Professor Xiang Peng, Shenzhen University with whom I have had the opportunity to interact and learn the subject.

I must thank Elsevier, OSA, and SPIE for allowing me to reuse some of their diagrams/images published in various journals. Thanks are due to Jessica Fricchione, Poppy Emerson, and Sarah Armstrong from IOP Publishing, UK, for guiding me on many issues while preparing the manuscript.

I wish to extend my sincere thanks to all my students, Dr Sudheesh K Rajput, Dr Isha Mehra, Dr Dhirendra Kumar, Dr Areeba Fatima, Alok K Gupta, Avishek Kumar, Praveen Kumar, and Yatish. My continuous interactions with them have led me to a deeper understanding of the subject. They helped me in preparing this manuscript and drawing some of the difficult diagrams.

Finally, I owe a lot to my family—particularly to my wife Rinki and my son Anvit for allowing me to spend long hours in preparing this manuscript and for their support throughout. I also thank my other family members in appreciation of their patience, encouragement, and constant support while this manuscript was being prepared.

Author biography

Naveen Kumar Nishchal

Dr Naveen Kumar Nishchal is an associate professor in the Department of Physics at the Indian Institute of Technology (IIT) Patna. He joined IIT Patna in December 2008. Dr Nishchal received his PhD degree in physics from IIT Delhi in 2005. He joined Instruments Research and Development Establishment, Dehradun, under Defence Research and Development Organisation, as a Scientist 'C' in July 2004 and worked until June 2007. Subsequently, he moved to IIT Guwahati and worked as an assistant professor in the Department of Physics from June 2007 to November 2008. He has been a visiting researcher to the Oulu Southern Institute, University of Oulu, Finland. His research interests include optical information processing, image encryption, watermarking, digital holography, interferometry, correlation-based optical pattern recognition, and fractional Fourier transform-based signal processing. Dr Nishchal is a senior member of OSA, SPIE, and life member of the Optical Society of India. He is a life member of Indian Science Congress Association, and Lasers and Spectroscopy Society of India. He has authored or co-authored 60 peer-reviewed international journal papers, two book chapters, and 150 papers in various conferences/seminars/symposia.

List of acronyms

AES	Advanced encryption standard
AT	Amplitude-truncated
BE	Beam expander
BR	Bacteriorhodopsin film
BS	Beam splitter
CC	Cross-correlation
CCA	Chosen-ciphertext attack
CCD	Charge-coupled device camera
CGH	Computer generated hologram
CMOS	Complementary metal-oxide semiconductor
COA	Ciphertext-only attack
CPA	Chosen-plaintext attack
CS	Compressive sensing
CT	Computed tomography
1D	One-dimensional
2D	Two-dimensional
3D	Three-dimensional
DCT	Discrete cosine transform
DES	Data encryption standard
DH	Digital holography
DOE	Diffractive optical element
DRPE	Double random phase encoding
DWT	Discrete wavelet transform
EFJPS	Encrypted fractional joint power spectrum
EMD	Equal modulus decomposition
ERA	Error reduction algorithm
FT	Fourier transform
FFT	Fast Fourier transform
FRT	Fractional Fourier transform
FrT	Fresnel transform
FWT	Fractional wavelet transform
GSA	Gerchberg–Saxton algorithm
GT	Gyrator transform
GWT	Gyrator wavelet transform
HIOA	Hybrid input–output algorithm
HM	Holographic mask
HOE	Holographic optical element
HT	Hartley transform
HWP	Half wave plate
IDEA	International data encryption algorithm
IFT	Inverse Fourier transform
JPS	Joint power spectrum
JTC	Joint transform correlator
KPA	Known-plaintext attack
LC	Liquid crystal
LCT	Linear canonical transform
LCTV	Liquid crystal television
LED	Light-emitting diode

MATLAB   Matrix laboratory
MEMS   Micro-electro-mechanical systems
MO   Microscope objective
MGSA   Modified Gerchberg–Saxton algorithm
MRI   Magnetic resonance imaging
MSE   Mean square error
MZI   Mach–Zehnder interferometer
NIST   National Institute of Standards and Technology
NPCR   Number of pixel change rate
OAC   Optical asymmetric cryptosystem
OD   Optical device
PCE   Peak-to-correlation energy
PCF   Phase contrast filter
PCI   Photon counting imaging
PD   Plastic diffuser
PI   Peak intensity
PK   Private key
POCSA   Projection-onto constraints sets algorithm
POF   Phase-only function
POM   Phase-only mask
PPM   Plasmonic phase mask
PRA   Phase retrieval algorithm
PRX   Photorefractive crystal
PSDOE   Polarization selective diffractive optical element
PSI   Phase-shifting interferometry
PSNR   Peak signal-to-noise ratio
PSR   Peak-to-sidelobe ratio
PT   Phase-truncated
PTFT   Phase-truncated Fourier transform
QPS   Quadratic phase system
QR   Quick response code
QWP   Quarter wave plate
RAM   Random amplitude mask
RE   Relative error
RGB   Red, green, blue
RP   Retardation plate
RPM   Random phase mask
RSAA   Rivest, Shamir, Adleman algorithm
SAA   Simulated annealing algorithm
SC   Symmetric cryptosystems
SHA   Secure hash algorithm
SLM   Spatial light modulator
SNR   Signal-to-noise ratio
SPM   Structured phase mask
SSE   Sum squared error
SWG   Subwavelength grating
UACI   Unified average change in intensity
VAR   Variance
VLC   VanderLugt correlator
WT   Wavelet transform
XOR   Exclusive OR

IOP Publishing

Optical Cryptosystems

Naveen K Nishchal

Chapter 1

Digital techniques of data and image encryption

1.1 Introduction

Information security is of paramount importance in today's digitally connected world. This is also called the *digital era*, in which the encryption is being considered as a fast-moving trend. Though advanced modern information security tools, storage, and retrieval mechanisms have been developed there are still enormous challenges posed by hacking tools, unsecure transmission channels, and ubiquity of the Internet. Therefore, there has been a rise in cyber security challenges globally, hence the users must be cyber prepared. Cyber security is impacting the industry. With the advent of advanced technologies such as *Internet-of-Things*, *Cloud Computing*, and *Artificial Intelligence*, it is envisaged that billions of devices would be connected. While such technologies provide several opportunities, they also pose threats to information security. Today most of the global web traffic is encrypted and it is expected that in future almost all the global web traffic will be fully encrypted. While this has enabled much greater privacy and helped prevent data breaches, cyber criminals are using these encrypted channels to propagate malware and exfiltrate data knowing that they can bypass traditional security inspection solutions that do not decrypt traffic [1–4].

The art and science of concealing information/data is called cryptography. The information/data/message to be concealed is called a plaintext (clear text) and the concealed form of message is called a ciphertext (encrypted text). In other words, cryptography is a process of converting plaintext into ciphertext and vice versa. The process of conversion from plaintext to ciphertext is called *encryption* and the reverse process that retrieves plaintext from ciphertext is called *decryption*. The ciphertext is a message that cannot be understood by anyone or is a meaningless message. A cipher is an algorithm used for encryption and decryption. The ciphertext is stored and transmitted to the intended user. The cryptography is not only used for protecting the information from theft or alteration but it is also used for user authentication [5–7].

doi:10.1088/978-0-7503-2220-1ch1

A cryptosystem, also referred to as a cipher system, is an implementation of cryptographic techniques and their accompanying infrastructure to provide information security services. Though cryptographic techniques have been in use for protecting information for thousands of years, the systematic study of cryptology as a science started around one hundred years ago. Therefore, cryptology is considered as a young science. Julius Caesar (around 100 BC) was known to use a form of encryption to convey secret messages to his Army Generals. The substitution cipher, known as the Caesar cipher is probably the most mentioned historic cipher in academic literature [3]. In this method, each character of a plaintext is substituted by another character to form the ciphertext. The variant used by Caesar was a shift by three ciphers. Each character was shifted by three places, so the character 'A' was replaced by 'D' and character 'B' was replaced by 'E' and so on. The characters would wrap around at the end, so character 'X' would be replaced by 'A'. An example of the character substitution based on Caesar's algorithm has been shown in figure 1.1.

Figure 1.2 shows the schematic of the modern encryption-decryption process. A plaintext is converted into a ciphertext through the encryption process, which upon use of correct keys returns the decrypted plaintext [4].

A basic cryptosystem has the following components [5]:
- Plaintext
- Encryption algorithm
- Ciphertext
- Decryption algorithm
- Encryption key
- Decryption key.

Figure 1.1. Example of character substitution based on Caesar's algorithm.

Figure 1.2. Encryption-decryption process.

A plaintext is converted into a ciphertext by applying the encryption algorithm and encryption key. The key space is a string of different keys that can be used to break the algorithm. It is generally accepted that a secure algorithm should use a key with length greater than 100 bits, because the number of bit permutation operations required to try 2^{100} keys is considered to be computationally infeasible for a conventional digital computing technique. A secure encryption algorithm is extremely sensitive to its keys. Various encryption algorithms have been developed and are being practiced. A ciphertext returns the plaintext only after use of the appropriate decryption algorithm and correct decryption key. A slight change to the keys would result in different ciphers. Thus for the successful retrieval of the plaintext, use of the correct decryption key and appropriate decryption algorithm is a must. In different types of cryptosystems, different encryption and decryption algorithms are used and correspondingly different encryption and decryption keys are generated.

While cryptography is the science of securing data, cryptanalysis is the science of analyzing and breaking a secure communication. The professionals involved in the process are called cryptanalysts. They are also called attackers. Attackers always wish to get the access of the encryption-decryption key so that plaintext can be retrieved. In classical cryptanalysis, several things are involved in the process, such as the interesting combination of analytical reasoning, the application of mathematical tools, pattern finding, patience, determination, and luck. With the passage of time, newer and reliable cryptosystems have been developed. On the other hand, attackers have also been creating improved logic to analyze the process to access the data. The pace of the development of information security technology is characterized by the creation of new methods and means of protection in the context of the storage, processing, and transmission of information. To date, much attention has been paid to the development of newer methods of intellectualization of various automated systems.

The cryptology embraces both cryptography and cryptanalysis. The cryptography can provide the following services [6].

- **Confidentiality (secrecy)**: it ensures that no one can read the concealed message except the authentic receiver. The data is kept secret from those who do not have proper credentials, even if that the data travels through an insecure medium.
- **Integrity (anti-tampering)**: it is assured that the authentic receiver has received message and it has not been altered in any way from the original.
- **Authentication**: it helps establish identity for authentication purposes. Actually, the process proves one's identity.
- **Non-repudiation**: it is a mechanism to prove that the sender really sent this message. Neither the sender nor the receiver can deny the transmission of the message.
- **Access control**: it requires that the access to information resources may be controlled by or for the authentic system.
- **Availability**: it requires that the system assets be available to authorized personnel, as and when needed.

1.2 Types of cryptography

Depending on the common uses, cryptography can be classified into two categories; *symmetric key cryptography* and *asymmetric key cryptography*. Symmetric key cryptography is a classical encryption method. It is referred to as a situation in which the key used for encryption is as used for decryption. In this case, key distribution must be performed prior to data transfer. Therefore, the security key plays a highly significant role because security directly depends on the nature of the key. Asymmetric key cryptography is an advanced encryption method. It is referred to as a situation in which the key used for encryption is different than the key used for decryption. In this case, a pair of keys, public and private keys, are used. The security is very high compared to the classical method of encryption.

Of late, hash functions are also considered as a type of cryptography, which establishes the authenticity of the user [7].

1.2.1 Symmetric key cryptography

Symmetric key cryptography, also known as secret key cryptography or conventional cryptography, refers to an encryption system in which the sender and receiver share a single common key that is used to encrypt and decrypt the message. The process is shown in figure 1.3. The used algorithm is known as the symmetric algorithm or secret key algorithm. The key is defined as a piece of information (a parameter) that determines the functional output of a cryptographic algorithm or cipher. The key used for encrypting and decrypting a message has to be known to all the authentic recipients or else the message could not be decrypted by conventional means [6]. The examples of symmetric key cryptography are discussed below.

- **Data encryption standard (DES)**: the DES was published in 1977 by the US National Bureau of Standards. It uses a 56-bit key and maps a 64-bit input block of plaintext onto a 64-bit output block of ciphertext. 56 bits is a rather small key for today's computing power.
- **Triple DES**: it is an improved version created after overcoming the shortcomings of DES. Since it is based on the DES algorithm, it is very easy to modify existing software to use Triple DES. It has the advantage of proven reliability and a longer key length that eliminates many of the shortcut attacks that can be used to reduce the amount of time it takes to break the DES.
- **Advanced encryption standard (AES)**: the AES is an encryption standard adopted by the US Government. The standard comprises three block ciphers, AES-128, AES-192, and AES-256. Each AES cipher has a 128-bit block size with key sizes of 128, 192, and 256 bits, respectively. The AES ciphers have been analyzed extensively and are now used worldwide.

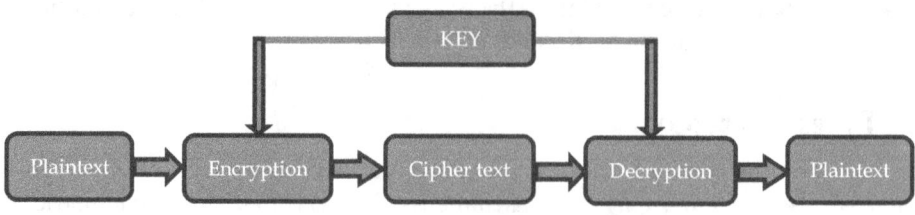

Figure 1.3. Symmetric key cryptography.

- **International data encryption algorithm (IDEA)**: the IDEA was developed in 1991. It uses a 128-bit key to encrypt a 64-bit block of plaintext into a 64-bit block of ciphertext. IDEA's general structure is very similar to DES. It performs 17 rounds, each round taking 64 bits of input to produce a 64-bit output, using per-round keys generated from the 128-bit key.

Key management in symmetric key systems

The symmetric key systems are simpler and faster but their main drawback is that the two parties must somehow exchange the key in a secure way and keep it secure after that. The key management caused a nightmare for the parties using the symmetric key cryptography. The worry was about how to get the keys safely and securely across all users so that the decryption of the message would be possible. This gave the chance for third parties to intercept the keys in transit to decode the secret messages. Thus, if the key was compromised, the entire coding system was compromised and a 'secret' would no longer remain a 'secret'.

1.2.2 Asymmetric key cryptography

Asymmetric key cryptography is also known as public key cryptography. It refers to a cryptographic algorithm which requires two separate keys, one of which is private and another is public. The public key is used to encrypt the message and the private one is used to decrypt the message. This method was developed to address the key management issue of symmetric key cryptography. The process of asymmetric cryptography is shown in figure 1.4. It is a very advanced form of cryptography. Officially, it was invented by Whitfield Diffie and Martin Hellman in 1975. The basic technique of public key cryptography was first discovered in 1973 by the British Clifford Cocks of Communications-Electronics Security Group but this was a secret until 1997. The examples of symmetric key cryptography are discussed below [6].

- **Digital signature standard (DSS)**: the DSS is a digital signature algorithm developed by the US National Security Agency to generate a digital signature for the authentication of electronic documents. DSS was put forth by the National Institute of Standards and Technology (NIST) in 1994.
- **RSA**: (Rivest, Shamir, and Adleman who first publicly described it in 1977) It is an algorithm for public-key cryptography. It is the first algorithm known to be suitable for signing as well as encryption, and one of the first great advances in public key cryptography. RSA is widely used in electronic

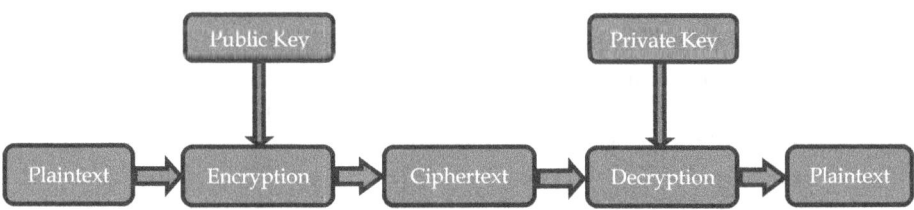

Figure 1.4. Asymmetric key cryptography.

commerce protocols, and is believed to be secure given sufficiently long keys and the use of up-to-date implementations.

- **ElGamal**: ElGamal is a public key method. It is used in both encryption and digital signing. The encryption algorithm is similar in nature to the Diffie–Hellman key agreement protocol and is used in many applications and uses discrete logarithms. ElGamal encryption is used in the free GNU Privacy Guard software.

1.2.3 Hash functions

A cryptographic hash function is a hash function that takes an arbitrary block of data and returns a fixed-size bit string, the cryptographic hash value such that any (accidental or intentional) change to the data will (with very high probability) change the hash value [7]. The data to be encoded is often called the message, and the hash values are sometimes called the message digest or simply digest. The ideal cryptographic hash function has four main properties:
- It is easy to compute the hash value for any given message.
- It is infeasible to generate a message that has a given hash.
- It is infeasible to modify a message without changing the hash.
- It is infeasible to find two different messages with the same hash.

The examples of hash functions are discussed below.
- **Secure hash algorithm (SHA)**: SHA hash functions are a set of cryptographic hash functions designed by the National Security Agency and published by the NIST as a US Federal Information Processing Standard. Because of the successful attacks on MD5, SHA-0 and theoretical attacks on SHA-1, NIST perceived a need for an alternative, dissimilar cryptographic hash, which became SHA-3. In October 2012, the NIST chose the Keccak algorithm as the new SHA-3 standard.

As multimedia, image, and video are becoming increasingly part of modern economy and social companions, ensuring security from malicious interference, theft, and unauthorized use has become the demand of the hour. Encryption of images is one of the well-known mechanisms to preserve confidentiality of images/ data over a reliable unrestricted public media, which is vulnerable to attacks. The image encryption algorithms can be classified into frequency-domain and spatial-domain algorithms. Both are able to protect the data/image with a high level of security. Their output encrypted images are either texture-like or noise-like images. From a security point of view, it is an obvious visual sign indicating the presence of an encrypted image that may contain some important information. It is apprehended that this will attract people's attention and can result in a significantly large number of attacks and analysis. The solution has been reported in the form that the original image is transformed into visually meaningful encrypted images. This is because people generally consider these images as normal images rather than encrypted ones.

Securing data/image is important in all the domains including medical diagnosis. There is a fear that patients' computed tomography (CT) and medical resonance imaging (MRI) scan results can easily be changed by hackers, thereby deceiving radiologists and artificial intelligence algorithms that diagnose malignant tumors. The hackers could access to add or remove medical conditions from the scans for the purpose of insurance fraud, ransom, and even homicide. A large number of techniques have been proposed in literature to date, each have an edge over the other, to catch up to the ever-growing need of security. The focus has been devising a mechanism for image encryption that should have the following characteristics.

• **Low correlation**	The value of correlation between the original and the encrypted image should be as low as possible. Ideally its value should be zero.
• **Large key space**	The key size should be very large since the more the key space, the higher the brute force search time would be.
• **Key sensitivity**	The image encryption algorithm should have high key sensitivity. In other words, a slight change in the key value should change the encrypted image significantly.
• **Entropy**	It is a measure of the degree of randomness or disorder. As the level of disorder rises, the entropy rises, and events become less predictable. The minimum entropy value should be zero and it happens when the image pixel value is constant in any location. The maximum value of entropy for an image depends on the number of gray scales. For an image with 256 gray scales, the maximum entropy is $\log 2(256) = 8$. The maximum value happens when all bins of the histogram have the same constant value, or, image intensity is uniformly distributed in [0,255].
Low time complexity	Usually, an encryption algorithm with high computational time is not recommended for practical applications. Therefore, an image encryption algorithm should have low time complexity.

The technology for information security using digital methods is being enhanced by applying more powerful algorithms. Longer key lengths are chosen such that current computers using the best cipher-cracking algorithms would require an unreasonable amount of time to break the key. When encryption key length becomes longer, the processing speed of digital techniques goes down. In order to counter the processing speed and security problem, in 1995 a new technology was proposed that used physical keys employing the principles of classical optics. Owing to the speed of light, it is envisaged that data can be secured at unparalleled speed along with parallel processing. Additionally, optics offers several degrees of freedom that could help encode information more securely [8–14]. Also, there is a natural match between optical processing for optical communications.

With the belief that cryptology based on the optics principle would provide a more complex environment and would be more resistant as compared to purely digital techniques, developing optical cryptosystems have gained much emphasis [13, 14]. Since 1995, a large number of research articles have appeared with so many different techniques. These topics are discussed in detail in the following chapters.

References

[1] Al Falou A (ed) 2018 *Advanced Secure Image Processing for Communications* (Bristol: IOP Publishing)

[2] Ramakrishnan S (ed) 2018 *Cryptographic and Information Security Approaches for Images and Videos* (Boca Raton, FL: CRC Press)

[3] Vacca J R (ed) 2017 *Computer and Information Security Handbook* 3rd edn (Amsterdam: Elsevier)

[4] Pfleeger C P, Pfleeger S L and Margulies J 2018 *Security in Computing* 5th edn (Noida: Pearson India)

[5] Beckett B 1988 *Introduction to Cryptology* (Oxford: Blackwell)

[6] Stallings W 2000 *Cryptography and Network Security; Principles and Practice* 2nd edn (Hoboken, NJ: Prentice Hall)

[7] van Tilborg H C A 2000 *Fundamentals of Cryptology* (Boston, MA: Kluwer)

[8] Naughton T J and Sheridan J 2005 Optics in information systems *SPIE Int. Tech. Gr. Newsletter* **16** 1–12

[9] Javidi B (ed) 2005 *Optical and Digital Techniques for Information Security* (Berlin: Springer)

[10] Javidi B (ed) 2006 *Optical Imaging Sensors and Systems for Homeland Security Applications* (New York: Springer)

[11] Alfalou A and Brosseau C 2009 Optical image compression and encryption methods *Adv. Opt. Photon* **1** 589–636

[12] Chen W, Javidi B and Chen X 2014 Advances in optical security systems *Adv. Opt. Photon* **6** 120–55

[13] Javidi B *et al* 2016 Roadmap on optical security *J. Opt.* **18** 083001

[14] Muniraj I and Sheridan J T 2019 *Optical Encryption and Decryption* (Bellingham, WA: SPIE Press)

IOP Publishing

Optical Cryptosystems

Naveen K Nishchal

Chapter 2

Optical techniques of image encryption: symmetric cryptosystems

2.1 Introduction

In an encryption system, an input data/image is encoded in such a fashion that only the application of the correct key would reveal the original information. The security of digital methods is being enhanced by using more powerful algorithms. Longer key lengths are chosen such that even advanced computers would require an unreasonable amount of time to break the key. Therefore, digital techniques are falling short of expectation due to the fact that when the encryption key length becomes longer, the processing speed goes down. It is primarily because digital techniques process data serially and in one dimension. Optical processing is inherently two-dimensional (2D) and does parallel processing. Every pixel of the 2D image can be both relayed and processed at the same time. So, when a large volume of data/information is to be processed, parallel processing offers enormous advantages. In addition to the fast speed, optical technology offers several advantages such as high space-bandwidth product, the possibility of including biometrics (face, fingerprint, iris, etc), and suitability for 2D data/images. Using the concept of holography, a three-dimensional (3D) object/scene can also be secured.

For encoding data securely, optics offers several degrees of freedom such as amplitude, phase, wavelength, polarization, spatial frequency, and optical angular momentum to encode data securely. An intensity sensing device, such as a charge-coupled device (CCD) camera cannot record any phase information. It is possible to tuck away an optically based message in only one small section of a 2D array, a trick that forces unauthorized users to find the message's position before they can begin to decode it. Therefore, it is believed that optical encryption techniques would provide a more complex environment and would be more resistant to attacks than purely electronic systems are.

doi:10.1088/978-0-7503-2220-1ch2

With the pioneering work on double random phase encoding (DRPE), optical techniques for information security have triggered much interest [1–3]. In the DRPE scheme, an input image to be encrypted is bonded with a random phase mask (RPM) and the product function is Fourier transformed. In the Fourier plane, another RPM is placed. Thus, the spectrum is again multiplied with the second RPM and its Fourier transformation is carried out, which gives a noisy image. This noisy image is called the encrypted image, which is a stationary white noise. Both the RPMs used in input and the Fourier plane are statistically independent and their values lie in the range [0,2π]. For decryption, the process is reversed and the conjugate of the RPMs at respective locations are used.

An optical information system consists of light source, lenses, mirrors, beam splitters, detectors, display devices such as a spatial light modulator (SLM), and a CCD camera. These components can be arranged in various configurations to suit the type of desired optical information processing setup.

Information in the form of a light wave passes through a converging lens that introduces delay or phase shift to the incident wavefront by an amount proportional to the thickness of lens, refractive index of the lens, and the wavelength of light. The light is distributed at the back focal plane of the lens, according to the spatial frequencies that are present in the original information. This spatial distribution in the back focal plane can be described mathematically as the Fourier transform of the input information. The Fourier transform capability of the converging lens is a crucial property of the optical information processors because it allows further manipulation of the optical information in the spatial frequency domain [4].

Holograms have been used in credit cards, identity cards, monetary bills, and many other important documents for security purposes. With the rapid technological advancement in the computers, CCD technology, image processing hardware and software, printers, scanners and copiers, it is increasingly becoming possible to reproduce complex pictures, symbols, logos, etc. Therefore, it has now become possible to duplicate a holographic pattern. The publication of pioneering work on DRPE has broadened the research area of information security like encryption, authentication, watermarking, and hiding. The optical encryption techniques have been realized and have stimulated research in the information security areas [5–8].

Pattern recognition is a science that concerns the description or classification of measurements. Pattern recognition techniques are often important components of intelligent systems and are used for both data processing and in decision-making. Opto-electronic techniques of pattern recognition for secure verification purposes are growing rapidly because of unique advantages offered by optical technologies [9]. Of late, the idea of authentication of credit cards, passports, driving license, and other personal identities has attracted much attention. The schemes use complex phase patterns that cannot be seen/copied by an intensity sensing device. Some security-enhanced optical security verification schemes have also been reported.

Amongst the optical security techniques, encryption is an effective approach to ensure the protection of information (text, image, and video) from unauthorized use or access. To secure the stored information, it is required to encrypt the data. An unauthorized user cannot reveal the original data without knowledge of the exact

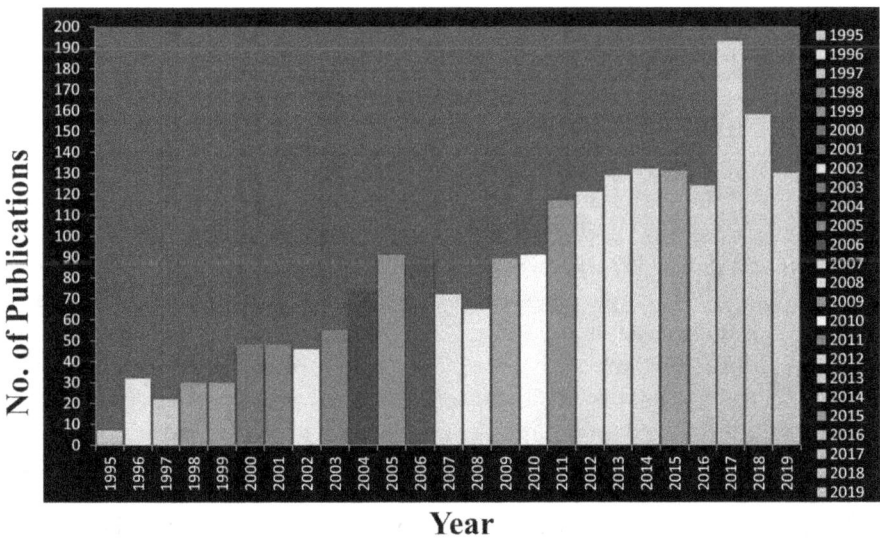

Figure 2.1. Literature growth on optical information security techniques.

random key code. Original information/data is encoded optically by using various encryption techniques including DRPE, polarization encryption, joint transform correlator (JTC)-based encryption, digital-optical image encryption, digital holography (DH)-based encryption, encryption using diffractive imaging, interference-based encryption, interference of polarized light-based encryption, chaos-based encryption, and quick response (QR) code-based encryption. Out of these, DRPE and its derivative techniques have been studied extensively. Figure 2.1 shows the growth in literature in the field of optical security over the years. There has been a constant increase in the number of articles published globally on the topic, which confirms the growing interest in the subject. No commercial device has been reported employing optical technology, but it is believed that such a product would be available in the coming future [8].

2.2 Encryption using linear canonical transforms

During the last few decades, many signal processing operations have been brought into the realm of Fourier optics. Some of them belong to the class of the linear canonical transforms (LCTs). LCT is a four parameter class of linear integral transforms, which is a flexible transform and possesses extra degrees of freedom without increasing the computational complexity [10]. It has received considerable attention over the period in optical information processing in general and information security applications in particular. Fractional Fourier transform (FRT), Fresnel transform (FrT), and gyrator transform (GT) belong to such a class while wavelet transformation, fractional convolution, and Wigner distribution belong to the category of linear combinations of LCTs or as cascades of such transformations. After the implementation of basic DRPE, subsequent optical encryption methods based on such transformations have been focused on encoding information. There are clear benefits (additional keys without extra computational cost) of these optical transforms in image encryption, watermarking, and steganography applications. The brief introduction of such systems has been discussed in the following subsections.

2.2.1 Double random phase encoding

In the DRPE technique, a primary image is encrypted using two RPMs, one bonded with the primary image and another placed in the Fourier domain, respectively. The schematic diagram of double random Fourier plane encoding is shown in figures 2.2(a) and (b). In DRPE, two statistically independent RPMs; $\exp\{i2\pi R_1(x,y)\}$ and $\exp\{i2\pi R_2(u,v)\}$ are employed at the image (input) and Fourier plane to encode an input image $f(x,y)$ into a ciphertext $E(x,y)$ as a complex-valued and stationary white noise. $R_1(x,y)$ and $R_2(u,v)$ are two independent white sequences uniformly distributed in [0,1].

The first step is to bond an input image $f(x,y)$ with an RPM, $\exp\{i2\pi R_1(x,y)\}$ and the combined function is Fourier transformed. The obtained expression is given as

$$E_1(u, v) = \iint [f(x, y) \times \exp\{i2\pi R_1(x, y)\}]\exp[-2\pi i(ux + vy)]\mathrm{d}x\mathrm{d}y \qquad (2.1)$$

Here, (x,y) and (u,v) denote the coordinates of image plane and Fourier plane, respectively. The second step is to bond the obtained Fourier spectrum with a statistically independent RPM, $\exp\{i2\pi R_2(u,v)\}$, and get the resultant again Fourier transformed. The second time computing Fourier transformation is also called obtaining inverse Fourier transformation. The obtained expression is given as

$$E(x, y) = \iint [E_1(u, v) \times \exp\{i2\pi R_2(u, v)\}]\exp[2\pi i(ux + vy)]\mathrm{d}u\mathrm{d}v \qquad (2.2)$$

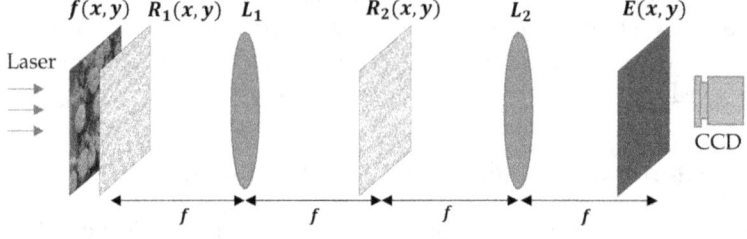

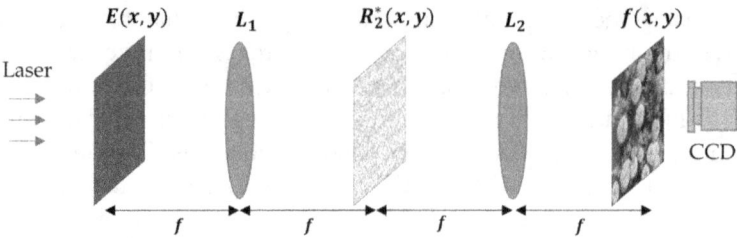

Figure 2.2. (a) Schematic diagram of the DRPE-based encryption scheme. (b) Schematic diagram of the DRPE-based decryption scheme.

The finally obtained expression, $E(x,y)$, is called the encrypted image. The decryption is the inverse of the encryption process, where all the operational steps described during encryption are performed in reverse. For successful decryption, there are two ways to follow. The first method is to use the conjugate of the respective RPMs in subsequent planes. In this case, the decryption process can be expressed as

$$E_1(u, v) = \Im[E(x, y)] \times \exp\{-i2\pi R_2(u, v)\} \tag{2.3}$$

$$f(x, y) = \Im^{-1}[E_1(u, v)] \times \exp\{-i2\pi R_1(x, y)\} \tag{2.4}$$

The symbols $\Im$ and $\Im^{-1}$ denote the Fourier transform and inverse Fourier transform operations, respectively. The second method is to use the conjugate of the encrypted image and respective original RPMs in subsequent planes. In this case, the decryption process can be expressed as

$$E_1(u, v) = \Im[\text{conj}\{E(x, y)\}] \times \exp\{i2\pi R_2(u, v)\} \tag{2.5}$$

$$f(x, y) = \Im^{-1}[E_1(u, v)] \times \exp\{i2\pi R_1(x, y)\} \tag{2.6}$$

It is difficult to generate the conjugate of the physical RPMs. Therefore, the use of the conjugate of the encrypted image is preferred, which can be easily generated through a four-wave mixing setup [11]. However, in the case of opto-electronic implementation through electrically addressed SLM, RPMs and their conjugates can be easily generated digitally and displayed. Another important issue to be discussed is that the use of RPM with the image to be encrypted in the input plane technically is not required for successful decryption. This is a drawback of the DRPE scheme as only Fourier-domain RPM is the required key for the successful retrieval of original data/information. MATLAB codes for a basic DRPE scheme have been given at the end of the chapter.

Statistical properties of the encoded image
It is important to note that the modulus of $\{f(x,y) \times \exp[i2\pi R_1(x,y)]\}$ is same as the modulus of $f(x,y)$. Therefore, the image is not encrypted in this case, although the RPM bonded input function $\{f(x,y) \times \exp[i2\pi R_1(x,y)]\}$ is a white noise [1]. This is demonstrated by evaluating the ensemble average of this input function on the random function $R_1(x,y)$:

$$<f(x, y) \exp[i2\pi R_1(x, y)]f(u, v) \exp[-i2\pi R_1(u, v)] > = f(x, y)f(u, v)\delta_{x-u}\delta_{y-v} \tag{2.7}$$

since $<\exp[i2\pi[R_1(x, y) - R_1(u, v)]]>_{R_1} = \delta_{x-u}\delta_{y-v}$ where δ_{x-u} is the Kronecker symbol. The symbol '<>' denotes the ensemble average. This white noise is nonstationary. If $f(x,y)$ is filtered with a phase-only filter of transfer function $\exp[i2\pi R_2(u,v)]$ and impulse response $h(x,y)$, then the obtained encrypted image is easy to decode.

In order to study the statistical properties of the encryption procedure, it is important to analyze the statistical property of the impulse response of a phase-only transfer function with a white noise. The following two properties are discussed.

Property 1: If $h(x,y)$ is the impulse response of a phase-only transfer function defined by $H(v, \mu) = \exp[i2\pi R_2(v, \mu)]$ where $R_2(v, \mu)$ is a white noise uniformly distributed in [0,1], then, for all x, y, u, v, ξ, η:

$$\langle h * (x - \xi, y - \eta)h(p - \xi, q - \eta)\rangle_b = \frac{1}{N^2}\delta_{x-p}\,\delta_{y-q} \qquad (2.8)$$

where * denotes the complex conjugate and

$$\delta_{x-p}\begin{cases}1 & \text{if } x - p = 0 \\ 0 & \text{otherwise}\end{cases} \qquad \delta_{y-q}\begin{cases}1 & \text{if } y - q = 0 \\ 0 & \text{otherwise}\end{cases}$$

Proof: For the proof of the property, the definition of Fourier transform of any function $h(x, y)$ can be used [3] and hence it is written as

$$h(x, y) = \frac{1}{N^2}\sum_{v=0}^{N-1}\sum_{\mu=0}^{N-1} \exp[i2\pi R_2(v, \mu)]\,\exp[i2\pi(vx + \mu y)] \qquad (2.9)$$

Thus the correlation of $h(x,y)$ is

$$\langle h * (x - \xi, y - \eta)h(p - \xi, q - \eta)\rangle_b = \frac{1}{N^4}\sum_{v=0}^{N-1}\sum_{v'=0}^{N-1}\sum_{\mu=0}^{N-1}\sum_{\mu'=0}^{N-1}$$
$$\times <\exp[i2\pi\{R_2(v', \mu') - R_2(v, \mu)\}]> R_2 \qquad (2.10)$$
$$\times \exp[i2\pi\{v'(p - \xi) + \mu'(q - \eta)\}$$
$$- \{v(x - \xi) + \mu(y - \eta)\}]$$

However, since $R_2(v, \mu)$ is a white noise uniformly distributed on [0,1],

$$<\exp[i2\pi\{R_2(v',\mu') - R_2(v, \mu)\}]>_{R_2} = \delta_{v-v}\,\delta_{\mu-\mu'} \qquad (2.11)$$

Substituting equation (2.11) in (2.10),

$$\langle h * (x - \xi, y - \eta)h(p - \xi, q - \eta)\rangle_{R_2}$$
$$= \frac{1}{N^4}\sum_{v=0}^{N-1}\sum_{\mu=0}^{N-1} \exp[i2\pi\{v(p - x) + \mu(q - y)\}] \qquad (2.12)$$

Applying the definition of discrete delta function,

$$\sum_{v=0}^{N-1}\sum_{\mu=0}^{N-1} \exp[i2\pi\{v(p-x)+\mu(q-y)\}] = \sum_{v=0}^{N-1} \exp[i2\pi v(p-x)] \sum_{\mu=0}^{N-1} \exp[i2\pi\mu(q-y)]$$

$$= N^2 \delta_{x-p}\delta_{y-q}$$

(2.13)

Thus, the property stated in equation (2.8) is obtained. This property proves that the impulse response of the function $\exp[i2\pi R_2(v, \mu)]$ is a stationary white noise.

Property 2: The encrypted function $\psi(x, y)$ is a stationary white noise with an autocorrelation function given as

$$<\psi * (x, y)\psi(\xi, \eta)> = \frac{1}{N^2}\sum_{u=0}^{N-1}\sum_{v=0}^{N-1} |f(u, v)|^2 \delta_{x-\xi}\delta_{y-\eta}$$

(2.14)

Proof:

$$\psi(x, y) = \sum_{u=0}^{N-1}\sum_{v=0}^{N-1} f(u, v)\exp[i2\pi R_1(u, v)h(x-u, y-v)]$$

(2.15)

Then,

$$<\psi * (x, y)\psi(\xi, \eta)> = \sum_{u=0}^{N-1}\sum_{v=0}^{N-1}\sum_{p=0}^{N-1}\sum_{q=0}^{N-1} f * (p, q)f(u, v)$$

$$\times <\exp[i2\pi\{R_1(p, q) - R_1(u, v)\}]$$

$$h * (x-p, y-q)h(\xi-u, \eta-v)>$$

(2.16)

However,

$$<\exp[i2\pi\{R_1(p, q) - R_1(u, v)\}]h * (x-p, y-q)h(\xi-u, \eta-v)>$$

$$= <\exp[i2\pi\{R_1(p, q) - R_1(u, v)\}]>_{R_1} \times <h * (x-p, y-q)h(\xi-u, \eta-v)>_{R_2}$$

(2.17)

$R_1(x, y)$ is a white noise uniformly distributed in [0,1], thus it can be written as

$$<\exp[i2\pi\{R_1(p, q) - R_1(u, v)\}]>_{R_1} = \delta_{u-p}\delta_{v-q}$$

(2.18)

Using **Property 1** and equation (2.18),

$$<\psi * (x, y)\psi(\xi, \eta)> = \sum_{u=0}^{N-1}\sum_{v=0}^{N-1}\sum_{p=0}^{N-1}\sum_{q=0}^{N-1} f * (p, q)f(u, v)$$

$$\times \delta_{u-p}\delta_{v-q} \times \frac{1}{N^2}\delta_{x-\xi}\delta_{y-\eta}$$

$$= \sum_{u=0}^{N-1}\sum_{v=0}^{N-1} f * (p, q)f(u, v)\frac{1}{N^2}\delta_{x-\xi}\delta_{y-\eta}$$

(2.19)

$$= \frac{1}{N^2}\sum_{u=0}^{N-1}\sum_{v=0}^{N-1} |f(u, v)|^2 \delta_{x-\xi}\delta_{y-\eta}$$

Property 2 helps establish the fact that although input plane RPM is not required for the retrieval of original information, the use of both the RPMs is important in converting the image to be encrypted into a white stationary noise. The use of input plane RPM makes the input image white but nonstationary and not encoded. Fourier plane RPM maintains whiteness and stationarizes and encodes the input image [1–3].

The publication of the pioneering 'double random phase encoding' attracted the attention of the research community and since then a large number of research papers have appeared in literature. This scheme has been implemented in different transform domains including fractional Fourier transform (FRT), Fresnel transform (FrT), gyrator transform (GT), wavelet transform (WT), and fractional Mellin transform. Also, other techniques such as asymmetric cryptosystems, phase-only encryption, multiple image encryption and color image encryption, have been reported. The brief introduction of some of the optical image encryption techniques in various transforms has been discussed in the following subsections.

2.2.2 Encryption using fractional Fourier transform

The fractional Fourier transform (FRT) is a generalization of the ordinary Fourier transform with an order parameter α. A Fourier transform is a first order FRT with $\alpha = 1$. The FRT operator follows the mathematical properties such as linearity, continuity, self-imaging, partial convolution/correlation. The properties and applications of the ordinary Fourier transform are special cases of those of the FRT [12–16]. In every area where Fourier transform and frequency domain concepts are used, the potential exists for generalizations and improvement by using the fractional transform. Efficient algorithms exist in literature to compute FRTs in about the same time as discrete Fourier transform [13–15]. Optical implementations based on bulk systems can be performed using either a single lens system or with a two lens system [12]. Therefore, the generalization of Fourier transform to the FRT comes at no additional cost whether computed digitally or implemented optically. If FRT of order α is denoted by $\mathfrak{I}^\alpha$, then its inverse is denoted as $\mathfrak{I}^{-\alpha}$. It has found many applications in optical and digital signal and image processing where the ordinary Fourier transform has traditionally played an important role. The FRT of a function $f(x_1, y_1)$ is defined as [16]

$$g(x_2, y_2) = \iint f(x_1, y_1) B_p(x_2, y_2; x_1, y_1) \mathrm{d}x_1 \mathrm{d}y_1 \tag{2.20}$$

where $B_p(\cdot)$ is the kernel of the 2D FRT given by:

$$B_p(x_2, y_2; x_1, y_1) = K \exp\left(i\pi \frac{x_2^2 + x_1^2 + y_2^2 + y_1^2}{\tan \alpha} - 2i\pi \frac{x_1 x_2 y_1 y_2}{\sin \alpha}\right) \tag{2.21}$$

Here, $\alpha = p\pi/2$ where p denotes order of the FRT and

$$K = \frac{\exp\left[-j\left(\frac{1}{4}\pi \, \text{sgn}(\sin \alpha)\right) - \frac{\alpha}{2}\right]}{|\sin \alpha|^{1/2}} \tag{2.22}$$

where K represents a complex constant and *sgn* refers to the *signum* function. For optical implementation of FRT, two geometries have been proposed; through a two lens system and a single lens system [12]. Figure 2.3 shows the schematic for obtaining FRT through a single lens system. In this case, the input function could be placed at any location within the focal length of the lens ($d < f$). The distance to input and output planes from the lens are the same but less than the focal length. The symbol 'f' denotes the focal length of the lens. The optical implementation of FRT with a single lens corresponds to free-space propagation (shearing of Wigner distribution function in the x-direction), passage through a lens (shearing in the y-direction), and again free-space propagation (shearing in the x-direction). The distance parameter, $d = f_1 \times \tan(\alpha/2)$, $f_1 = f \times \tan(\alpha/2)$, where f is the focal length of the lens.

The ciphertext generated by using DRPE in the FRT domain is written as [17–20]

$$E(x, y) = \mathfrak{F}^{\beta}\{\mathfrak{F}^{\alpha}\{[f(x, y) \times \exp(i2\pi R_1(x, y))] \times \exp(i2\pi R_2(u, v))\}\} \tag{2.23}$$

Here, values of α and β are important for the retrieval of original information in addition to the use of respective RPMs. For decryption, the usual reverse process of encryption, as explained in section 2.2.1, is to be followed. Because the order of the FRT can take any value between 0 and 1 therefore, if the DRPE is implemented in the FRT domain the information of orders enlarges the key space and the cryptosystem becomes stronger. For this reason, fractional Fourier domain encoding attracted the attention of the researchers community.

Extending the concept of FRT, extended FRT has been reported [21]. The benefit of the transform is that the distance to input and output planes from the lens can be different, less than or greater than the focal length of the lens. Hence, the distances become asymmetric and offer additional degrees of freedom while applying it for encryption applications.

Figure 2.4 shows the schematic for obtaining extended FRT through a single lens system. In this case, the input function could be placed at any location, which can be

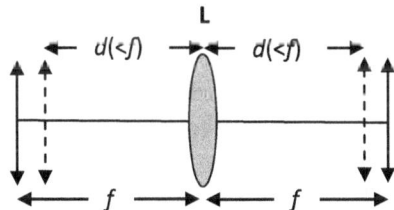

Figure 2.3. Schematic diagram for optical implementation of FRT with a single lens system.

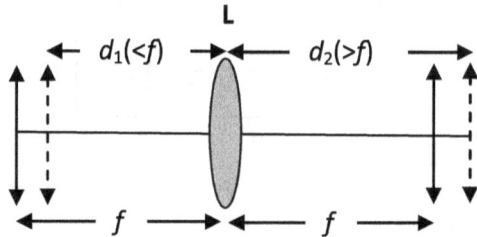

Figure 2.4. Schematic diagram for optical implementation of extended FRT with a single lens system.

less than or greater than the focal length of the lens and this fractional focal plane can be arbitrarily decided. Following 1D representation for convenience, the extended FRT of a function $f(x)$ is defined as [21],

$$g(u) = K \int f(x) \exp\left(i\pi \frac{a^2 x^2 + b^2 u^2}{\tan \alpha} - 2i\pi \frac{abxu}{\sin \alpha}\right) dx \qquad (2.24)$$

The function $g(u)$ is related to $f(x)$ by an FRT with three parameters a, α, and b. K is a complex constant. The parameters a, α, and b are called quadratic phase system (QPS) parameters and in general they are complex quantities. Performing an extended FRT on a function is equivalent to expanding the function 'a' times, performing an FRT of order α, and contracting the resultant distribution 'b' times. The QPS parameters are related to the distances d_1 and d_2 and the focal length f of the lens through the following expressions.

$$a^2 = \frac{1}{\lambda} \frac{\sqrt{f - d_2}}{\sqrt{f - d_1}} \frac{1}{[f^2 - (f - d_1)(f - d_2)]^{1/2}} \qquad (2.25)$$

$$\alpha = \arccos\left[\frac{\sqrt{f - d_1}\sqrt{f - d_2}}{f}\right] \qquad (2.26)$$

$$b^2 = \frac{1}{\lambda} \frac{\sqrt{f - d_1}}{\sqrt{f - d_2}} \frac{1}{[f^2 - (f - d_1)(f - d_2)]^{1/2}} \qquad (2.27)$$

If the distance parameters d_1 and d_2 are taken as the same and less than the focal length of the lens, then the expression reduces to FRT. When the distance parameters are considered larger than the focal length of the lens then QPS parameters become complex quantities. Thus, while implementing extended FRT for image encryption, four additional keys (QPS and wavelength) are generated. This enhances the key space and hence the security [22].

Figure 2.5 shows the schematic for extended FRT domain DRPE for image encryption employing RPMs and FRT parameters as keys. It is important to note that optical implementation of the scheme does not demand any extra components but at the same time helps enhance the security many times. This aspect is very

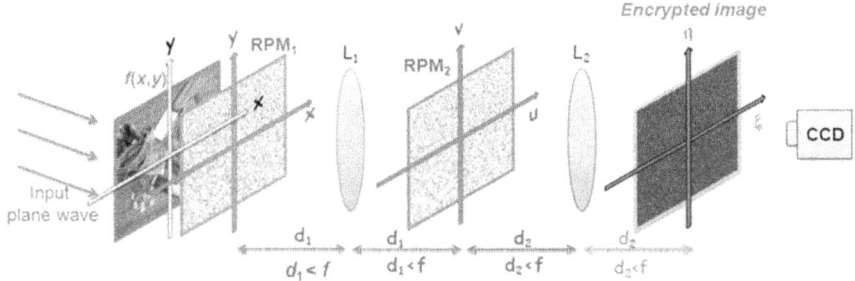

Figure 2.5. Schematic diagram of the FRT domain DRPE-based encryption scheme.

important for making a practical cryptosystem [22]. A MATLAB code has been given at the end of the chapter.

2.2.3 Encryption using Fresnel transform

Securing information under the DRPE framework in Fresnel transform (FrT) domain has been reported in literature [23, 24]. In FrT-based image encryption techniques, optical wavelength, propagation distance, and sampling parameters are considered as additional keys. Thus, key space is enlarged and hence security of such systems becomes stronger. FrT-based schemes are so strong that even if there is small change in any of the parameters, such as wavelength or propagation distance, the original image is not retrieved. Various types of multiplexing (rotation, position, wavelength) schemes have been implemented in the FrT domain in order to secure multiple images.

In the FrT domain DRPE technique, a primary image to be encrypted is bonded with an RPM and is Fresnel transformed. The obtained spectrum is modulated with another RPM and is again Fresnel transformed, which results in an encrypted image. Both the RPMs are statistically independent. The schematic diagram of the DRPE scheme in the FrT domain is shown in figure 2.6.

Mathematically, FrT is computed through the Fresnel-Kirchhoff formula. The FrT of a function $f(x,y)$ is written as [4],

$$F(u, v) = \mathfrak{S}_\lambda^d[f(x, y)] = \frac{\exp\left\{\dfrac{i2\pi d}{\lambda}\right\}}{\sqrt{i\lambda d}} \iint f(x, y)$$
$$\times \exp\left[\frac{i\pi}{\lambda d}((x - u)^2 + (y - v)^2)\right]dxdy \qquad (2.28)$$

where $\mathfrak{S}_\lambda^d$ denotes the FrT operation, d denotes the propagation distance, λ is the optical wavelength, and (x,y) and (u,v) represent the coordinates of input and output domains, respectively. The ciphertext generated by using DRPE in the FrT domain is written as

$$E(x, y) = \mathfrak{S}_\lambda^{d_2}\left\{\mathfrak{S}_\lambda^{d_1}\{[f(x, y) \times \exp(i2\pi R_1(x, y))] \times \exp(i2\pi R_2(u, v))\}\right\} \qquad (2.29)$$

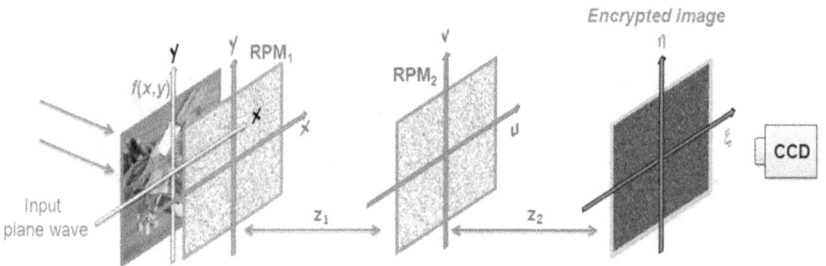

Figure 2.6. Schematic diagram of the FrT domain DRPE-based encryption scheme.

Here, values of d_1, d_2, and λ are important for successful retrieval of the original information in addition to respective RPMs. For decryption, the usual reverse process of encryption, as explained in section 2.2.1, is to be followed.

2.2.4 Encryption using gyrator transform

The gyrator transform (GT) is a linear canonical integral transform, which produces the rotation in twisted position-spatial frequency planes of phase space [25]. Similar to FRT, gyrator transform is also a generalization of the ordinary Fourier transform with a parameter α. For $\alpha = 0$, it corresponds to identity transform and for $\alpha = \pi/2$, it corresponds to Fourier transform. The gyrator transform is periodic and additive with respect to parameter α.

Similar to FRT, gyrator transform has been used in image encryption applications [26, 27]. This is because of parameter α, which connects with the angle of gyrator transform and provides additional security to the encryption scheme. This is also optically implemented employing cylindrical lenses. Mathematically, GT of any function $f(x,y)$ is defined as [25],

$$g(x_2, y_2) = \frac{1}{|\sin \alpha|} \iint f(x_1, y_1) \exp\left(\frac{i2\pi(x_2y_2 + x_1y_1)\cos \alpha - (x_1y_2 + x_2y_1)}{\sin \alpha}\right) dx_1 dy_1 \quad (2.30)$$

Here, (x_1, y_1) are the co-ordinates of the input function and (x_2, y_2) are the co-ordinates in the gyrator domain and α is the angle of the GT. The ciphertext generated by using DRPE in the gyrator domain is then written as:

$$E(x, y) = GT^\beta\{GT^\alpha\{[f(x, y) * \exp(i2\pi R_1(x, y))] * \exp(i2\pi R_2(u, v))\}\} \quad (2.31)$$

Here, $GT^\alpha\{.\}$ and $GT^\beta\{.\}$ represent the gyrator transform operations applied for angles α and β, respectively. The functions $R_1(x, y)$ and $R_2(u, v)$ are two random phase value distribution, lying in the interval [0,1].

2.2.5 Encryption using wavelet transform

Wavelet transform is a signal processing tool used for the analysis of optical and digital signals. It has good local optimization features as well as the multi-resolution analysis features, which makes it suitable for information processing applications [28]. Discrete wavelet transform (DWT) is any wavelet transform that uses a discrete set of wavelet scales and transformation to process signals. WT has been used in analyzing and processing optical signals due to its scaling and shift parameters. Its application in image compression is well established. Therefore, there is an inherent capability of combining WT not only in securing information but in compressing the secured data. This helps communicate the secured data through conventional communication channels. The optical implementation of the WT for 2D objects has been reported. Considering 1D representation, a mother wavelet $h(x)$ is a finite-duration window function that can generate a family of daughter wavelets by varying scale a and shift b.

$$h_{a,b}(x) = \frac{1}{\sqrt{a}}h\left(\frac{x-b}{a}\right) \tag{2.32}$$

The mother wavelet should satisfy the admissibility conditions that it must be oscillatory, have fast decay to zero, and integrate to zero. WT is defined as an inner product between a signal and a set of wavelets as

$$w_f(a,b) = \int_{-\infty}^{\infty} h_{a,b}^*(x)f(x)\mathrm{d}x \tag{2.33}$$

where * denotes the complex conjugate and $w_f(a,b)$ is considered as a function of spatial shift b for each fixed scale a that displays the information $f(x)$ at various resolution levels.

Further, the scaling factor of WT and fractional orders of FRT has been combined, which is called fractional WT (FWT). This scheme has been used for image encryption. The FWT is a linear transformation without introducing any cross-term interference [29, 30]. It combines the virtues of the WT as well as the FRT and possesses multi-resolution features. The concept behind this transform is extracting the fractional spectrum of the signal via FRT operation followed by WT operation of the fractional spectrum. This concept of FWT has also been applied to optical information security applications. The fractional orders of FRT and scaling factors of wavelets serve as the additional keys to the cryptosystem. If these keys are chosen randomly then the unauthorized people cannot retrieve the actual input data. So the information can be well protected.

The use of WT in information security is more suitable while particularly securing multispectral data [31, 32]. In which, images collected through different sensors at different wavelengths are received. Such multi-spectral images need to be fused (merging of different frequency sub-bands) first and then secured. The multispectral data received from satellites and airborne sensors are becoming increasingly available for analysis for various applications such as remote sensing. Fusion techniques play an important role in processing all such multispectral data. This technique maximizes the information of the fused image and thus improves the image content. If such fusion schemes can be applied for information security application, then the degrees of freedom may be greatly enhanced. Combining the concept of wavelets with GT, gyrator wavelet transform (GWT) has also been reported, which finds applications in image encryption and compression [33, 34].

2.2.6 Encryption using cosine transform

The discrete cosine transform (DCT) expresses a finite sequence of data points in terms of cosine functions at different frequencies. Since fewer cosine functions are required to approximate a typical signal, DCT finds application in data/image compression. This property attracted the use of DCT and discrete fractional cosine transform in image encryption applications. The 2D DCT of an image $f(x,y)$ of size $M \times N$ is defined as [35]

$$B(\xi, \eta) = \frac{2}{\sqrt{LK}}\gamma(\xi)\gamma(\eta) \sum_{u=0}^{L-1}\sum_{v=0}^{K-1} A(u, v)\cos\left[\frac{(2u + 1)\xi\pi}{2L}\right]\cos\left[\frac{(2v + 1)\eta\pi}{2K}\right] \quad (2.34)$$

The corresponding inverse DCT is expressed as

$$A(u, v) = \frac{2}{\sqrt{LK}} \sum_{\xi=0}^{L-1}\sum_{\eta=0}^{K-1} \gamma(\xi)\gamma(\eta)B(\xi, \eta)\cos\left[\frac{(2u + 1)\xi\pi}{2L}\right]\cos\left[\frac{(2v + 1)\eta\pi}{2K}\right] \quad (2.35)$$

$$\gamma(\xi) = \begin{cases} \dfrac{1}{\sqrt{2}} & \text{for } \xi = 0 \\ 1 & \text{for } \xi = 1, 2, \ldots, L - 1 \end{cases},$$

$$\gamma(\eta) = \begin{cases} \dfrac{1}{\sqrt{2}} & \text{for } \eta = 0 \\ 1 & \text{for } \eta = 1, 2, \ldots, K - 1 \end{cases} \quad (2.36)$$

The discrete fractional cosine transform has a similar relationship with the discrete FRT. In this case, fractional order is an added property to the encryption scheme.

2.2.7 Encryption using fractional Mellin transform

Mellin transform is closely related to the Laplace and Fourier transforms and exhibits a certain invariance to object magnification. Considering 1D form, the Mellin transform of a function $f(x)$ is defined by [4]

$$M(s) = \int_0^\infty f(x)x^{s-1}\,\mathrm{d}x \quad (2.37)$$

where, in the most general case, s is a complex variable. Considering the complex variable $s = i2\pi f_X$ and substituting $x = e^{-\xi}$, the Mellin transform is expressed as

$$M(i2\pi f_X) = \int_{-\infty}^\infty f(e^{-\xi})e^{-i2\pi f_X\xi}\,\mathrm{d}\xi \quad (2.38)$$

Equation (2.34) represents the Fourier transform of the function $f(e^{-\xi})$. The Mellin transform can be implemented with an optical Fourier transforming system. The most important property of this transform is that its magnitude is independent of scale-size changes in the input. This property has attracted the use of the Mellin transform to image processing applications including image encryption [36]. Similar to the concept of FRT, fractional Mellin transform has also been used in developing a nonlinear encryption scheme to survive the conventional attacks. Also, extending the properties of FRT, fractional Hartley transform has been proposed for image encryption [37]. In all the cases, the basic DRPE framework has been combined with other optically implementable transforms to enhance the level of security. The attributes of transforms provide attractive features in security applications.

MATLAB codes

I. Fourier transform domain encryption

```
%Double Random Phase encoding (DRPE)
%PT is the plaintext
%Phase_mask1 and Phase_mask2 are the two random phase
  masks to be used as the keys.
PT = imread('D:\Program\images\godisgreat.bmp');
    %reading the image to be encrypted
PT = double(PT(:,:,1));
PT = PT./max(max(PT));
figure;imagesc(abs(PT));colormap(gray);
    title('original input image');
%defining the two random phase masks for the two keys
[M,N]= size(PT);
phase_values1 = rand(M);
phase_values2 = rand(M);
phase_mask1 = exp(j*2*pi*phase_values1);%First Key
phase_mask2 = exp(j*2*pi*phase_values2);%Second Key
%%%%%Encryption
%First Fourier transform
A=PT.*phase_mask1;
A=frt2(A);
%Second Fourier transform
B=A.*phase_mask2;
B=fft2(B);%ciphertext
figure;imagesc(abs(B));colormap(gray);
    title('encrypted image');
```

```
%%%%%decryption
D=ifft2(ifft2(B).*conj(phase_mask2));
```
figure;*imagesc*(*abs*(*D*));colormap(gray);
```
  title('decrypted image');
```

II. Fractional Fourier transform

function[*X*]=*frt*(*obj,p*)
```
% MATLAB code for fractional Fourier transform
% obj is the input subjected to fractional Fourier
  transform
% p is the order of fractional Fourier transform
```
N=size(*obj*);
```
%p=0.5;
```
n=N(1);
[*mx,my*]=*meshgrid*(−*n*/2:1:*n*/2−1);
```
% Action of first lens
```
L1=exp(−*j**(*pi/n*)* *tan*(*p*pi*/4)*(*mx.*mx+my.*my*));
*A=obj.*L1;*
```
% Free space propagation
```
Af=fft2(*A*);
L2 = exp(−*j**(*pi/n*)* *sin*(*p*pi/2*)*(*mx.*mx+my.*my*));
*B=Af.*L2*;
C=ifft2(*B*);
```
% Action of second lens
```
L3=exp(−*j**(*pi/n*)* *tan*(*p*pi/4*)*(*mx.*mx+my.*my*));
*D=C.*L3;*
```
%Multiplication by a phase factor
```
pf=exp(−*j**(*pi/n*)).*/ sqrt*(*abs*(*sin*(*p*pi/2*)));
*X=D.*pf;*

III. Fractional Fourier transform domain encryption

```
% Double Random Phase encoding (DRPE) in fractional
  Fourier domain
% The code uses the function frt to perform the fractional
  Fourier transform.
% This function requires two inputs: the entity to be
  fractional
% Fourier transformed and the order of transformation.
```

```
% PT is the plaintext
% Phase_mask1 and Phase_mask2 are the two random phase
  masks to be used as the keys.
```
$PT = imread('D:\Program\images\godisgreat.bmp');$
```
    %reading the image to be encrypted
```
$PT = doublc(PT(:,:,1));$
$PT = PT./max(max(PT));$
$figure; imagesc(abs(PT)); colormap(gray);$
```
    title('original input image');
%defining the two random phase masks for the two keys
```
$[M,N]= size(PT);$
$phase\_values1=rand(M);$
$phase\_values2=rand(M);$
$phase\_mask1=exp(j*2*pi*phase\_values1);$%First Key
$phase\_mask2=exp(j*2*pi*phase\_values2);$%Second Key
```
%%%%Encryption
%First Fractional Fourier transform with fractional
    order 0.25
```
$A=PT.*phase\_mask1;$
$A=frt(A,0.25);$
```
%Second Fractional Fourier transform with fractional
    order 0.45
```
$B=A.*phase\_mask2;$
$B=frt(B,0.45);$%ciphertext
$figure; imagesc(abs(B));$ colormap(gray);
```
    title('encrypted image');
%%%%decryption
```
$D=frt(frt(B,-0.45).*conj(phase\_mask2),-0.25);$
$figure; imagesc(abs(D));$ colormap(gray);
title('decrypted image');

IV. Gyrator transform

```
functionqt=gyrator(q,a)
% Matlab code for fast algorithm of discrete gyrator
  transform
%qis an input signal and a is rotation angle
% Direct DGT
```

$[M,N] = size(q);$

$mm = ((0: M-1) - (M)/2)/sqrt(M);$

$nn = ((0: N-1) - (N)/2)/sqrt(N);$

$[x,y] = meshgrid(nn, fliplr(mm));$

$[u,v] = meshgrid(mm, fliplr(nn));$

$p1 = exp(-2*j*pi*x.*y*tan(a/2));$

$p2 = fftshift(exp(-2*j*pi*u.*v*sin(a)));$

$qt = p1.*(ifft2(fft2(p1.*q).*p2));$

```
end
```

References

[1] Refregier P and Javidi B 1995 Optical image encryption using input plane and Fourier plane random encoding *Proc. SPIE* **2565** 62–8

[2] Refregier P and Javidi B 1995 Optical image encryption based on input plane encoding and Fourier plane random encoding *Opt. Lett.* **20** 767–9

[3] Javidi B and Ahouzi E 1998 Optical security system with Fourier plane encoding *Appl. Opt.* **3** 6247–55

[4] Goodman J W 2007 *Introduction to Fourier Optics* 3rd edn (New Delhi: Viva Books)

[5] Alfalou A and Brosseau C 2009 Optical image compression and encryption methods *Adv. Opt. Photon.* **1** 589–636

[6] Liu S, Guo C and Sheridan J T 2014 A review of optical image encryption techniques *Opt. Laser Technol.* **57** 327–42

[7] Chen W, Javidi B and Chen X 2014 Advances in optical security systems *Adv. Opt. Phot.* **6** 120–55

[8] Javidi B *et al* 2016 Roadmap on optical security *J. Opt.* **18** 083001

[9] Javidi B and Horner J L 1994 Optical pattern recognition for validation and security verification *Opt. Eng.* **33** 1752–6

[10] Healy J J, Kutay M A, Ozaktas H M and Sheridan J T (ed) 2015 *Linear Canonical Transform: Theory and Applications* (Berlin: Springer)

[11] Unnikrishnan G, Joseph J and Singh K 1998 Optical encryption system that uses phase conjugation in a photorefractive crystal *Appl. Opt.* **37** 8181–5

[12] Lohmann A W 1993 Image rotation, Wigner rotation, and the fractional Fourier transform *J. Opt. Soc. Am.* A **10** 2181–6

[13] Ozaktas H M, Arikan O, Kutay M A and Bozdagi G 1996 Digital computation of the fractional Fourier transform *IEEE Trans. Signal Process.* **44** 2141–50

[14] Garcia J, Mas D and Dorsch R G 1996 Fractional Fourier transform calculation through the fast Fourier transform algorithm *Appl. Opt.* **35** 7013–8

[15] Khan G S, Nishchal N K, Jospeh J and Singh K 2001 Fractional Fourier transform and its applications: a bibliographic review *Asian J. Phys.* **10** 251–99

[16] Ozaktas H M, Zalevsky Z and Kutay M A 2001 *The Fractional Fourier Transform with Applications in Optics and Signal Processing* (Chichester: Wiley)

[17] Unnikrishnan G and Singh K 2000 Double random fractional Fourier-domain encoding for optical security *Opt. Eng.* **39** 2853–9

[18] Unnikrishnan G, Joseph J and Singh K 2000 Optical encryption by double-random phase encoding in the fractional Fourier domain *Opt. Lett.* **25** 887–9

[19] Unnikrishnan G, Joseph J and Singh K 2001 Fractional Fourier domain encrypted holographic memory by use of an anamorphic optical system *Appl. Opt.* **40** 299–306

[20] Unnikrishnan G and Singh K 2001 Optical encryption using quadratic phase systems *Opt. Commun.* **193** 51–67

[21] Hua J, Liu L and Li G 1997 Extended fractional Fourier transforms *J. Opt. Soc. Am.* A **14** 3316–22

[22] Nishchal N K, Joseph J and Singh K 2003 Optical encryption using cascaded extended fractional Fourier transform *Opt. Memory Neural Net.* **12** 139–45

[23] Situ G and Zhang J 2004 Double random phase encoding in the Fresnel domain *Opt. Lett.* **29** 1584–6

[24] Shi Y, Situ G and Zhang J 2007 Multiple-image hiding in the Fresnel domain *Opt. Lett.* **32** 1914–6

[25] Rodrigo J A, Alieva T and Calvo M L 2006 Optical system design for orthosymplectic transformations in phase space *J. Opt. Soc. Am.* A **23** 2494–500

[26] Rodrigo J A, Alieva T and Calvo M L 2007 Applications of gyrator transform for image processing *Opt. Commun.* **278** 279–84

[27] Singh H, Yadav A K, Vashisth S and Singh K 2014 Fully phase image encryption using double random-structured phase masks in gyrator domain *Appl. Opt.* **53** 6472–81

[28] Mallat S 2008 *A Wavelet Tour of Signal Processing* 3rd edn (New York: Academic)

[29] Chen L and Zhao D 2005 Optical image encryption based on fractional wavelet transform *Opt. Commun.* **254** 361–7

[30] Vilardy J M, Useche J, Torres C O and Mattos L 2011 Image encryption using the fractional wavelet transform *J. Phys.: Conf. Ser.* **274** 012047

[31] Mehra I and Nishchal N K 2014 Image fusion using wavelet transform and its application to asymmetric cryptosystem and hiding *Opt. Express* **22** 5474–82

[32] Mehra I and Nishchal N K 2015 Wavelet-based image fusion for securing multiple images through asymmetric keys *Opt. Commun.* **335** 153–60

[33] Mehra I and Nishchal N K 2015 Optical asymmetric image encryption using gyrator wavelet transform *Opt. Commun.* **354** 344–52

[34] Mehra I, Fatima A and Nishchal N K 2018 Gyrator wavelet transform *IET Image Process* **12** 432–7

[35] Zhou N, Wang Y and Gong L 2011 Novel optical image encryption scheme based on fractional Mellin transform *Opt. Commun.* **284** 3234–42

[36] Meng X F, Cai L Z, Yang X L, Xu X F, Dong G Y, Shen X X, Zhang H and Wang Y R 2007 Digital color image watermarking based on phase-shifting interferometry and neighboring pixel value subtraction algorithm in the discrete-cosine-transform domain *Appl. Opt.* **46** 4694–701

[37] Singh P, Yadav A K and Singh K 2017 Phase image encryption in the fractional Hartley domain using Arnold transform and singular value decomposition *Opt. Lasers Eng.* **91** 187–95

IOP Publishing

Optical Cryptosystems

Naveen K Nishchal

Chapter 3

Fully-phase image encryption

3.1 Introduction

A *phase object* is defined as a completely transparent object but has an optical thickness that varies from point to point. Such an object introduces phase difference between disturbances that pass through different parts of it. Consequently, the disturbances immediately behind the object and in the conjugate image plane produce the same amplitude at all points but show variations in phase from point to point. The human eye is sensitive to intensity only and cannot detect phase changes. Therefore, the field of view appears uniformly bright. The variations in optical thickness in the object cause variations in intensity in the image so that the phase object is rendered visible [1].

Imaging of phase objects has been a subject of considerable interest in the field of optics. In microscopy, there are several objects, which are largely transparent, thus absorbing little or no light. When light passes through such an object, the predominant effect is the generation of a spatially varying phase shift. For a number of applications, the spatial distribution of the phase is the only available and/or the only desirable information. In the ideal case, the phase object is absolutely invisible. Various techniques have been applied for the imaging and visualization of phase objects. The Zernike's phase contrast method is a well-established technique for the visualization of phase perturbations. A fabricated phase contrast filter (PCF) is used for imaging the phase objects [2].

The use of phase has a longstanding history in optical image processing and is usually correlated with the use of coherent illumination. Invariably, a 2D image is encoded in the phase of an optical wavefront and decoded back again after processing using the phase contrast technique. The commercial availability of SLMs has widened the scope of complex-valued representation for computation applications. Phase encoding offers advantages in terms of computational efficiencies or light power efficiencies [3]. The architectures employing phase is regarded as

doi:10.1088/978-0-7503-2220-1ch3

analog optical processing architectures. In this case, each scalar is encoded in a quantized phase value.

In the DRPE architecture, the input images to be encrypted are usually intensity representations. It was proposed that if the input images are represented as phase-only versions, then fully-phase encryption can be considered as a nonlinear encryption and the cryptosystem would provide higher security than linear (amplitude) encryption. It was shown that a fully-phase encryption performs better than amplitude-based encryption when the system bandwidth is limited by a moderate amount. The better performance is observed in the presence of additive noise with respect to the mean square error (MSE) [4–7]. The MSE metric measures the square of the Euclidean distance between two vectors and is mathematically tractable. The detailed mathematical derivation for the theoretical estimate of MSEs for the fully-phase decryption method has been given in [5]. The problem with the fully phase encryption arises only during the decryption process that the information content of the decrypted phase image cannot be read by any intensity detector. Therefore, a technique is required for converting the phase image into an amplitude image. This additional process enhances the level of security and adds complexity to the optical system. The optical decryption can be implemented in the common path interferometer setup by the use of a single phase or, if desired, a combined phase key.

3.2 Phase imaging

The phase contrast method is a technique for converting a spatial phase modulation into a spatial intensity modulation. Frits Zernike proposed the phase contrast technique based on the spatial filtering principle [2]. The method has the advantage that the observed intensity is linearly related to the phase-shift introduced by the object. Consider a transparent object with amplitude transmittance,

$$t_A(x,y) = \exp[i\varphi(x,y)] \tag{3.1}$$

Expanding $\exp[i\varphi(x,y)]$, the following expression is obtained.

$$e^{i\varphi(x,y)} = 1 + i\varphi(x,y) - \frac{1}{2}\varphi^2(x,y) - \frac{1}{6}i\varphi^3(x,y) + \frac{1}{24}\varphi^4(x,y) + \dots \tag{3.2}$$

The structure of a phase object is normally invisible because the irradiance distribution is constant. The basic purpose of phase contrast imaging is to allow visualization of such phase objects. The transparent object is illuminated by a coherent beam of light in an image forming setup.

For mathematical simplicity, it is assumed that the object has a magnitude of unity and the finite extent of the entrance and exit pupils is neglected. Also, there is a necessary condition to achieve linearity between phase-shift and intensity. The condition is that the variable part of the object-induced phase-shift, $\Delta\varphi(x,y)$, should be small compared with 2π radians. In this case, the crudest approximation to amplitude transmittance is [8],

$$t_A(x,y) = e^{i\varphi_0}e^{i\Delta\varphi} \approx e^{i\varphi_0}[1 + i\Delta\varphi] \tag{3.3}$$

The higher orders have been neglected. The image produced can be written as

$$I \approx |1 + i\Delta\varphi(x,y)|^2 \approx 1 \tag{3.4}$$

The diffracted light arising from the phase structure is not observable in the image plane because it is in phase quadrature with the strong background. Zernike proposed that if this phase-quadrature relation could be modified then the two terms might interfere and produce observable variations of the image intensity. The background is focused on the optical axis in the focal plane but the diffracted light arising from higher spatial frequencies is spread away from the optical axis. Therefore, Zernike proposed that a phase-changing plate, also called PCF, can be inserted in the focal plane to modify the phase relation between the focused and the diffracted light [2].

A PCF or phase-changing plate consists of a glass substrate on which a small transparent dielectric dot is deposited. The dot is centered on the optical axis in the focal plane and has a thickness and index of refraction such that it should change the phase of the focused light by either $\pi/2$ radians or $3\pi/2$ radians relative to the phase retardation of the diffracted light. If the phase retardation is by $\pi/2$ radians, the intensity in the image plane becomes

$$I \approx |\exp[i(\pi/2)] + i\Delta\varphi(x,y)|^2 = |i\{1 + \Delta\varphi(x,y)\}|^2$$
$$\approx 1 + 2\Delta\varphi(x,y) \tag{3.5}$$

The image intensity has become linearly related to the variations of phase-shift $\Delta f(x,y)$. This situation is referred to as positive phase contrast. If the phase retardation is by $3\pi/2$ radians, the intensity in the image plane becomes

$$I \approx |\exp[i(3\pi/2)] + i\Delta\varphi(x,y)|^2 = |-i\{1 - \Delta\varphi(x,y)\}|^2$$
$$\approx 1 - 2\Delta\varphi(x,y) \tag{3.6}$$

This case is referred to as negative phase contrast. The basic scheme for phase contrast is a 4-f imaging setup as shown in figure 3.1.

Extending the concept of Zernike's technique of phase contrast into the domain of full range $[0,2\pi]$ phase modulation, a generalized phase contrast was proposed [9]. The method is based on the application of a spatial filtering and phase shifting

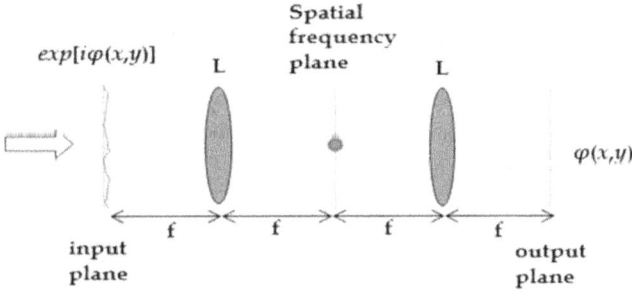

Figure 3.1. Schematic for phase contrast imaging.

operation in the Fourier plane of an optical system. The generalized technique is not restricted to the small-scale phase regime of the Zernike's method. Higher order terms in the expansion (equation (3.2)) are taken into account. In the generalized method, the phase-changing plate introduces π phase difference between the background and the higher spatial frequencies. In a general 4-f imaging setup, as shown in figure 3.1, the second lens performs a Fourier transform of the π phase shifted and the non-phase shifted light. The interference between these components generates intensity distribution in the output plane, which can be visualized, photographed, or grabbed by a CCD camera. In short, the system behaves like a common path interferometer in which the low-frequency of the information, encoded in the phase of the incident wavefront, is phase-shifted in the Fourier plane.

3.3 Fully-phase encryption

Let (x,y) denote the space coordinates, and (u,v) the coordinates in the Fourier domain. The real-valued function $f(x,y)$ denotes a primary 2D image to be encrypted, and $E(x,y)$ the encrypted image. The fully-phase encryption of the input image $f(x,y)$ is done in three steps. Firstly, the input image is phase-encoded, which can be mathematically expressed as $\exp[if(x,y)]$. The second step is multiplying the phase-encoded image with an RPM. Finally, the product is convolved by a function, which is the impulse response of a phase-only transfer function. The encryption method can be implemented either optically or electronically.

The phase-encoded image $\exp[if(x,y)]$ is multiplied by an RPM, $\exp[i2\pi R_1(x,y)]$ and its Fourier transformation is obtained. Thus, generated Fourier spectrum is again modulated by another RPM, $\exp[i2\pi R_2(u,v)]$ and the output is inverse Fourier transformed. $R_1(x,y)$ and $R_2(u,v)$ are two independent white sequences uniformly distributed in [0,1]. The obtained encrypted image can be written as

$$E(x,y) = \Im^{-1}\{\Im\{\exp[if(x,y)] \times \exp[i2\pi R_1(x,y)]\} \\ \times \exp[i2\pi R_2(u,v)]\}$$
(3.7)

The RPM, $\exp[i2\pi R_1(x,y)]$ converts phase-encoded image $\exp[if(x,y)]$ into white but nonstationary noise, whereas the use of second RPM, $\exp[i2\pi R_2(u,v)]$ encrypts the image into a stationary white noise. The mathematical details for DRPE approach have been discussed in chapter 2.

The decryption process is the inverse of the encryption process, where all the operational steps described in the encryption stage are performed in reverse. For successful decryption, there are two ways to follow. The first method is to use the conjugate of the respective RPMs in subsequent planes. The second method is to use the conjugate of the encrypted image and respective RPMs in subsequent planes. Considering the use of the conjugate of the encrypted image, the decryption process can be written as

$$\exp[if(x,y)] = \Im\{\Im^{-1}\{[\text{conj}\{E(x,y)\}] \times \exp[i2\pi R_2(u,v)]\} \\ \times \exp[i2\pi R_1(x,y)]\}$$
(3.8)

Thus, through the inverse process of encryption and using the correct RPMs in respective positions, the phase-encoded image is retrieved successfully. The next task is to convert the decrypted phase image into an intensity image. For this purpose, the obtained phase-encoded image is Fourier transformed. In the Fourier plane, a prefabricated PCF of appropriate dimension is placed. The zero order of diffraction is modulated with the phase filter and obtaining one more Fourier transform results in the desired intensity image. The schematic for fully-phase encryption is given in figure 3.2. The basic setup is based on holography and four-wave mixing. In DRPE-based fully-phase encryption, the usual procedure of the DRPE method is followed. The encrypted image can be recorded holographically in a recording material such as a photorefractive crystal/photopolymer. The encrypted information is stored in a photorefractive crystal and the phase-encoded data is retrieved by a phase conjugate read-out scheme and the same two RPMs are used during the encryption process. The phase conjugation technique is used for generating the conjugate of the encrypted image through the four-wave mixing setup [10]. The phase conjugation process corrects the phase distortions mostly of aberrations of optical components. The original data is recovered with an interferometer.

An expanded coherent beam of light is split into two parts: reflected and transmitted beams. The reflected beam is made incident on an SLM, which encodes the desired amplitude pattern/image into the phase component of the incident wavefront at the object plane. It is to be noted that the SLM should be operating in fully-phase mode with 2π phase modulation. The input RPM can be placed just after the SLM or a digitally generated RPM can be multiplied with the phase-encoded image, which can be displayed onto the SLM. The SLM is placed in the front focal plane of a Fourier transforming lens. Thus, the phase-encoded input image modulated with the RPM is Fourier transformed. The spectrum is then modulated with another RPM and one more Fourier transformation is obtained, which gives

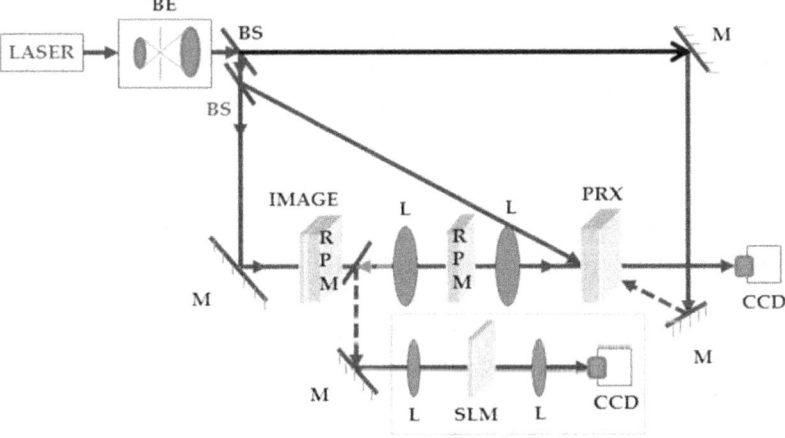

Figure 3.2. Schematic diagram for optically implementing fully phase encryption. BE: beam expander, BS: beam splitter, M: mirror, RPM: random phase mask, L: lens, PRX: photorefractive crystal, SLM: spatial light modulator, CCD: charge-coupled device camera.

the encrypted image. This encrypted optical beam interferes with the transmitted beam and the interference pattern is recorded into a photorefractive crystal. To record the encrypted image, the shutter is closed so that the readout beam would not erase the recorded hologram. The angle between the two beams is kept optimum for better signal-to-noise ratio (SNR).

For decryption, the photorefractive crystal is illuminated with a conjugate of the reference (transmitted) beam thus generating the conjugate of the encrypted image through the four-wave mixing principle. The recorded encrypted image hologram is readout by opening the shutter to obtain the decrypted image through phase conjugation. This generated beam traverses the path followed during the encryption process. Hence the random phase values introduced during encryption get cancelled and the phase-encoded image is retrieved. Now this phase-encoded pattern has to be converted into an intensity pattern. For this conversion, an extra optical setup is attached, as shown in figure 3.2, within the dotted line. The phase pattern is Fourier transformed and the spectrum is filtered with the phase-changing plate, which upon one more Fourier transformation results in the desired intensity image. An important point to note here is that for successful decryption both the RPMs in the respective locations are required. As has been seen in DRPE, while using an intensity image for encryption both the RPMs are required but for decryption only the Fourier domain RPM is needed. Hence, a fully-phase encryption offers higher security. In this case, it is also possible that the encryption process may use only one RPM and still the security will not be compromised.

For generating a phase contrast image, various techniques have been implemented. Phase contrast using optically addressed liquid crystal SLM [11] and bacteriorhodopsin (BR) film [12] have also been reported. One disadvantage of optically addressed liquid crystal SLM and BR film-based phase contrast techniques is that these techniques are intensity-dependent. Therefore, higher intensity orders also introduce phase shifts, which are undesirable. Also the phase shift introduced is object-dependent. Phase-only optical encryption and decryption with a readout based on the generalized phase-contrast method has also been reported, which uses a common path interferometer configuration [13, 14]. The design of the PCF depends on the operating wavelength and the diameter of the fabricated hole. The hole is fabricated usually in a $\lambda/2$ thick layer (for introducing π phase shift) of thin film coating (matching wavelength) upon a glass optical flat. The coating material should be effectively nonabsorbing at the operating wavelength [14]. This technique provides a simple and robust architecture, as it uses only two phase masks. But for generating a phase contrast image, this requires extremely precise alignment.

An optical encryption system based on the encryption of information using the phase component of a wavefront has been reported [15]. The method utilized the polarization for the visualization of decoded information through two parallel aligned liquid crystal SLMs. But this was suitable for binary optical phase information. Further, to achieve the phase-only optical encryption, the principle of interference has been applied [16]. In this case, use of a reference beam reduced the alignment requirement and pixel-to-pixel mapping problem. The original image is reconstructed by the interference of a reference wave with only two RPMs without any decryption filter. The method gave emphasis on image decryption.

For a practical system, it is desired to use commercially available devices. Considering this goal, it was demonstrated that phase contrast can be obtained by using an electrically addressed SLM as a PCF [17]. This alleviated the need of fabrication of a PCF. The property of an electrically addressed liquid crystal SLM can be utilized to introduce the required phase-shift corresponding to a phase-encoded image displayed on the SLM. For making a digital PCF, a white blank pattern of size of the dimension of the SLM, in which a black square of the size of the required hole in the centre is digitally created. This blank pattern with a dark dot of required size is displayed on the SLM. The white pattern would give a $\pi/2$ or π phase-shifts with respect to the dark pattern and corresponds to the phase shifter mask of a conventional PCF. The SLM working in the fully-phase mode can introduce $\pi/2$ or π phase-shifts between the lower and higher spatial frequencies. Thus, SLM can be used for employing the conventional phase contrast or generalized phase contrast technique. Use of an SLM in the filter plane does not require the manual fabrication of a phase contrast filter and it offers the freedom of the generation of the phase contrast filter of different phase shifts and dimensions. In the classical phase contrast technique, it is possible to improve the image contrast by making the phase-shifting dot partially absorbing [8]. This requirement can also be fulfilled by the use of a liquid crystal SLM with appropriate modulation.

Further, considering the benefits (additional degrees of freedom) of fractional Fourier domain encoding, fully-phase encryption has been demonstrated [18–20]. A phase image obtained from an amplitude image using extended FRT was encrypted in the framework of DRPE. Figure 3.3 shows the schematic for extended FRT based fully-phase encryption. A quadratic phase system (QPS) offers a continuum of planes, in which encoding can be done. Exact information of the encoding plane is required for the retrieval of image/data. The decryption is the inverse process of encryption.

An input amplitude image is first phase-encoded and displayed on an SLM. Input domain RPM is placed close to the SLM and the function is fractional Fourier transformed. Again in the fractional domain another RPM is placed, which further

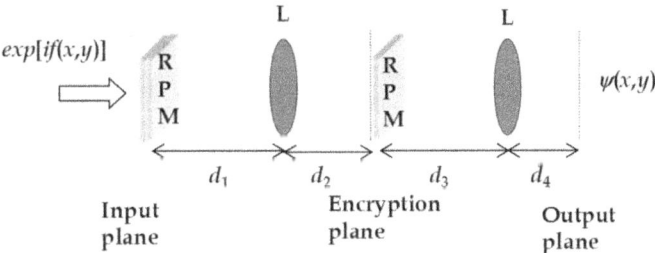

Figure 3.3. Schematic for extended FRT domain fully phase encryption. RPM: random phase mask, L: lens, d_1, d_2, d_3, d_4: distance parameters for defining the quadratic phase system parameters for extended FRT operation.

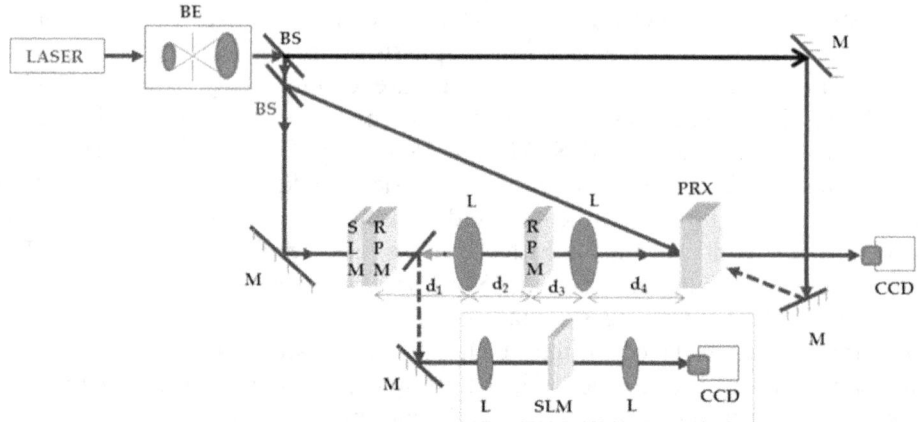

Figure 3.4. Schematic for extended FRT domain fully phase encryption. BE: beam expander, BS: beam splitter, M: mirror, RPM: random phase mask, L: lens, PRX: photorefractive crystal, SLM: spatial light modulator, d_1, d_2, d_3, d_4: different fractional Fourier planes, CCD: charge-coupled device camera.

modulates the function and gives an encrypted image, which is recorded holographically into a photorefractive crystal. The experimental setup is shown in figure 3.4. The recorded hologram is readout by opening the shutter to obtain the phase conjugate encrypted image. This generated beam traverses the path followed during the encryption process and after getting the introduced random phase values cancelled, the phase-encoded image is retrieved. Through the phase contrast technique applied using an SLM, the phase image is converted into an amplitude image. The optical setup for phase contrast has been shown within the dotted line. The experimental results of fully-phase encryption employing DRPE in the fractional domain have been shown in figure 3.5 [18].

Figure 3.5(a) shows the amplitude image used for encryption, which is phase-encoded and displayed onto an SLM working in fully-phase mode. The encrypted image formed at the photorefractive material plane is imaged onto a CCD and is shown in figure 3.5(b). The decrypted image obtained after using correct QPS parameters and the same RPM, and after using phase contrast is shown in figure 3.5(c). The decrypted image obtained after using the correct QPS parameters but different RPM is shown in figure 3.5(d). The image when it is decrypted with a set of QPS parameters ($a = 37.8 - 37.8i$, $b = 17.9 + 17.9i$, $\alpha = 1.57 - 0.23i$, $c = 10.84$, $d = 32.06$, and $\beta = 1.76$, corresponding to $f_1 = 13.5$ cm, $f_2 = 7.5$ cm, $d_1 = 15$ cm, $d_2 = 6.9$ cm, $d_3 = 11.9$ cm, and $d_4 = 8$ cm, respectively), which are different from those used for recording (but with the same RPMs), is as shown in figure 3.5(e) [18].

It can be seen that, even if same RPMs are used the original image cannot be retrieved, if correct QPS parameters are not used. In addition to all these important parameters, the decoded phase image cannot be read unless a technique is used for its conversion into intensity representation. The decrypted image with correct QPS parameters and the same RPM, but without using phase contrast, is shown in figure 3.5(f).

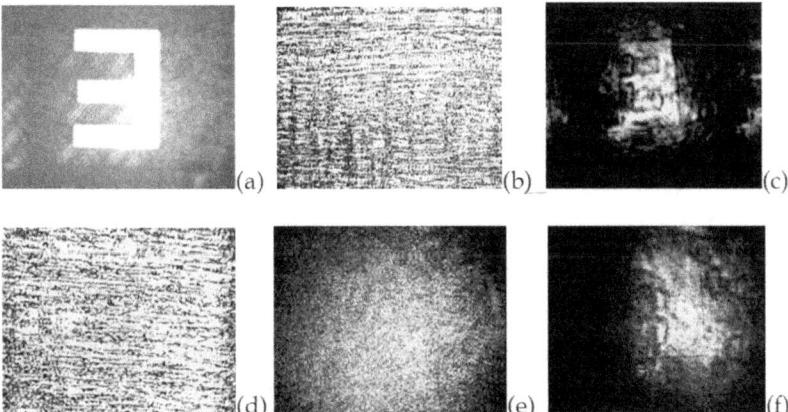

Figure 3.5. Experimental results of fully-phase encryption using fractional domain random phase encoding. (a) Original amplitude image to be encrypted, (b) encrypted image with QPS parameters ($a = 37.6 - 37.6i$, $b = 18 + 18i$, $\alpha = 1.57 - 0.23i$, $c = 29.2$, $d = 87.5$, and $\beta = 1.77$), (c) decrypted image obtained after applying correct QPS parameters and the same RPMs and after using phase contrast technique, (d) decrypted image with correct QPS parameters and different RPM and after using phase contrast technique, (e) decrypted image with different QPS parameters ($a = 37.8 - 37.8i$, $b = 17.9i + 17.9i$, $\alpha = 1.57 - 0.23i$, $c = 10.84$, $d = 32.06$, and $\beta = 1.76$) and the same RPM and after using phase contrast technique, and (f) decrypted image with correct QPS parameters and the same RPM and without using the phase contrast technique. Reproduced with permission from [18].

Since security of an encryption technique depends on the size of the key used, an enlarged key space provides enhanced security. In the amplitude-based encryption techniques using FRT, the two RPMs and the QPS parameters constitute the keys. For decryption, only the conjugate of the fractional plane RPM along with the correct QPS parameters are required. The input plane RPM associated with the amplitude object gets automatically canceled when recorded with the intensity detector like the CCD camera. In contrast, in fully-phase encryption, the input image is a phase image. Hence, both the RPMs constitute the keys, and are essentially required for decryption along with the QPS parameters. In addition, a method for converting phase image into an amplitude image is required. Thus, the security in phase encryption using extended FRT is higher as compared to amplitude-based encryption [17–19]. Hence, the idea was further extended to cascading of extended FRTs in order to use more than two RPMs, to expand the key space [20]. Photorefractive materials have also been reported to be used as a PCF that also alleviates the need of fabrication [21]. To obtain a stationary phase contrast image, the crystal must be illuminated for a long time to let the refractive index change in the crystal reach saturation. For real-time implementation of PCF, a photorefractive material with fast response should be employed. The chief application of such an optical experiment was reported for secure data storage (holographic memory) and the key component in the experimental work is a dynamic phase-only SLM.

References

[1] Longhurst R S 1986 *Geometrical and Physical Optics* 3rd edn (London: Orient Longman)

[2] Zernike F 1955 How I discovered phase contrast *Science* **121** 345–9

[3] Naughton T J 2010 Phase in optical image processing *AIP Confer. Proc.* **1236** 235–40

[4] Neto L G 1998 Implementation of image encryption using the phase-contrast technique *Proc. SPIE* **3386** 284–90

[5] Towghi N, Javidi B and Luo Z 1999 Fully phase encrypted image processor *J. Opt. Soc. Am.* A **16** 1915–27

[6] Tan X, Matoba O, Shimura T, Kuroda K and Javidi B 2000 Secure optical storage that uses fully phase encryption *Appl. Opt.* **39** 6689–94

[7] Javidi B, Towghi N, Maghzi N and Verrall S C 2000 Error reduction techniques and error analysis for fully phase and amplitude based encryption *Appl. Opt.* **39** 4117–30

[8] Goodman J W 2007 *Introduction to Fourier Optics* 3rd edn (New Delhi: Viva Books)

[9] Gluckstad J 1996 Phase contrast image synthesis *Opt. Commun.* **130** 225–30

[10] Unnikrishnan G, Joseph J and Singh K 1998 Optical encryption system that uses phase conjugation in a photorefractive crystal *Appl. Opt.* **37** 8181–5

[11] Vorontsov M A, Justh E W and Beresnev L A 2001 Adaptive optics with advanced phase contrast techniques I. High-resolution wave front sensing *J. Opt. Soc. Am.* A **18** 1289–99

[12] Castillo M D I, Sanchez-de-la-Llave D, Garcia R R, Olivos-Perez L I, Gonzalez L A and Rodriguez-Ortiz M 2001 Real-time self-induced nonlinear optical Zernike type filter in a bacteriorhodopsin film *Opt. Eng.* **40** 2367–8

[13] Mogensen P C and Gluckstad J 2000 Phase-only optical encryption *Opt. Lett.* **25** 566–8

[14] Mogensen P C and Gluckstad J 2001 Phase-only optical decryption of a fixed mask *Appl. Opt.* **40** 1226–35

[15] Mogensen P C and Gluckstad J 2000 A phase image optical encryption system with polarization encoding *Opt. Commun.* **173** 177–83

[16] Seo D H and Kim S J 2003 Interferometric phase-only optical encryption system that uses a reference wave *Opt. Lett.* **28** 304–6

[17] Nishchal N K, Joseph J and Singh K 2003 Fully phase encryption by phase contrast using electrically addressed spatial light modulator *Opt. Commun.* **217** 117–22

[18] Nishchal N K, Joseph J and Singh K 2003 Fully phase encryption using fractional Fourier transform *Opt. Eng.* **42** 1583–8

[19] Nishchal N K, Joseph J and Singh K 2004 Fully phase-based encryption using fractional order Fourier domain random phase encoding: error analysis *Opt. Eng.* **43** 2266–73

[20] Nishchal N K, Joseph J and Singh K 2004 Fully phase encrypted memory using cascaded extended fractional Fourier transform *Opt. Lasers Eng.* **42** 141–51

[21] Liu J, Xu J, Zhang G and Liu S 1995 Phase contrast using photorefractive LiNbO$_3$:Fe crystals *Appl. Opt.* **34** 4972–5

IOP Publishing

Optical Cryptosystems

Naveen K Nishchal

Chapter 4

Joint transform correlator-based schemes for security and authentication

4.1 Introduction

Pattern recognition is defined as the categorization of input data into identifiable classes. In other words, pattern recognition classifies an observed data into one of the previously determined classes. It is achieved through the extraction of relevant features or attributes of a data set from a background of irrelevant details. It finds applications across several disciplines, such as image analysis, image and speech recognition, nondestructive testing, missile guidance, and tactile sensing. Pattern recognition is a computationally demanding problem. Because of parallel processing capabilities and the potential to carry out computations at the speed of light, optical processors are considered as an alternative tool for implementing some of the necessary operations.

Correlation is an important tool in determining whether the input function matches a stored function. The optical correlation between two functions, $t(x,y)$ and $r(x,y)$, is written as

$$c(x,y) = t(x,y) \otimes r(x,y) \tag{4.1}$$

where the symbol $\otimes$ represents the correlation operation. Employing Fourier transforms, equation (4.1) is rewritten as

$$\Im\{c(x,y)\} = T(f_x, f_y) \times R^*(f_x, f_y) \tag{4.2}$$

where $T(f_x, f_y)$ denotes the Fourier transform of $t(x,y)$ and $R^*(f_x, f_y)$ is the complex conjugate of the Fourier transform of $r(x,y)$. The convolution of two functions in space domain is equivalent to the operation of multiplying their individual transform and inverse transforming. There are two well-known optical architectures for finding the correlation between two functions, namely reference and target functions; the VanderLugt correlator (VLC) [1] and joint transform correlator (JTC) [2].

doi:10.1088/978-0-7503-2220-1ch4

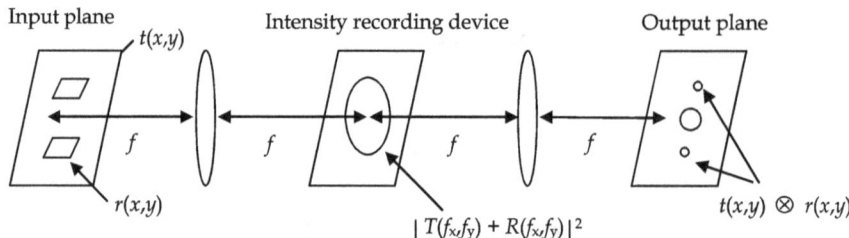

Figure 4.1. A typical JTC layout.

The VLC architecture utilizes a frequency-plane mask, which is also called a matched filter. This filter is the complex conjugate of the Fourier transform of any function. This mask (matched filter) is synthesized holographically and can effectively control both the amplitude and phase of a transfer function. In this geometry, an input function, say target $t(x,y)$, is first Fourier transformed and in the frequency plane the matched filter $R^*(f_x, f_y)$ is placed. Illuminated by a coherent beam of light, the input target's frequency spectrum $T(f_x, f_y)$ is multiplied with the matched filter and one more Fourier transform operation is carried out, which results in an auto-correlation peak, provided there is a match between the target and the reference function. The VanderLugt filter is very sensitive to the exact position of the frequency-plane mask and synthesizing a matched filter is a tedious and time-consuming task.

A typical JTC layout has been shown in figure 4.1. In this architecture, the target and reference functions are simultaneously placed side-by-side in the input plane. This joint function is Fourier transformed and the power spectrum is recorded using an intensity-sensing device. The power spectrum, which is known as the joint power spectrum (JPS), is fed to the input of the optical processor and its Fourier transformation is obtained to achieve the correlation outputs. The correlation plane contains three spatially separated terms: two auto-correlation peaks and a strong undesired zero order (dc), if the reference function is matched with the target function.

The JTC architecture has some advantages over the VLC geometry, such as that the precise alignment of the filter transparency is not required and it can be used in real-time systems. The price paid for the JTC geometry is the reduction of the space-bandwidth product of the input transducer. Also, a square-law device is required for the detection and display of the intermediate intensity pattern.

4.2 DRPE using JTC

In the DRPE scheme, for retrieval of the encrypted data, the conjugate of RPMs is required. Generating the conjugate of RPMs is possible through holography, which is a cumbersome task, therefore the JTC approach has been considered as an alternative technique. In this method, the generation of the conjugate of RPMs is not required, rather the same RPMs being used. The idea is that instead of auto-correlation peaks resulting in the output plane, when the reference matches with the target; the original images are reproduced if the correct RPMs are used during the decryption process.

In the JTC-based scheme, an input image to be encrypted is bonded with a phase mask (RPM1) and is placed side-by-side with a key phase mask (RPM2). The joint input image is Fourier transformed and its power spectrum is recorded with a camera, which is called the encrypted data. Due to the joint transform architecture, the alignment of the image and the key is robust as compared to the DRPE architecture. The JTC scheme has been used under the DRPE framework for image encryption-decryption applications. In addition to that, it has also been used for authentication applications, in which decrypted images do not appear in the output plane but autocorrelation peaks appear, which proves the authenticity.

The pioneering work reported in 1994 for validation and security verification opened the window for the variety of modifications in JTC with several newer applications [3]. The basic idea was to permanently bond an RPM to a primary identification amplitude pattern (e.g. a photograph, fingerprint, picture of a face, picture of the iris, biometric signature, personal identification number), which cannot be seen or copied by a camera. The phase mask being a phase-only function is invisible under ordinary light. In a different approach, a 1D space integrating optical processor has been reported, which can perform position-invariant 2D image correlation by 1-D scanning. The processor can be used for verifying the authenticity of images [4]. Further, an incoherent optical correlator based on phase-only filters has been proposed for security applications. The phase encoding was combined with the method of generalized projections onto constraint sets (POCS) implemented by an iterative Fourier transform algorithm [5]. A nonlinear JTC with improved performance measures was implemented for security and validation application [6]. To authenticate an ID card or other object, a specially encoded hologram was attached. A nonlinear material, bacteriorhodopsin (BR) was used in the JTC architecture. Correlations were made between holograms of an RPM recorded either in silver halide or a photopolymer material.

The term image encryption and DRPE were combined for the first time in the year 2000 with JTC [7]. In this system, the encrypted JPS was recorded by a camera. The process alleviated the need of displaying the complex encrypted data. Also, the scheme does not require the conjugate of the Fourier plane RPM for decryption. The system was implemented with photorefractive-based optical correlation geometry. The schematic of the DRPE using JTC system is depicted in figure 4.2. Figure 4.2(a) shows the encryption process, in which the input image to be encrypted is bonded with an RPM and the key code, another RPM is placed alongside the input image. This forms the joint input image. The joint input images can be placed in vertical directions or horizontal directions or at any angular positions. The minimum distance between the two images is governed by the basic JTC theory [2]. The joint images are Fourier transformed and the JPS, which is called the encrypted image or encrypted power spectrum, is recorded using a CCD camera. As the RPM has the complex value, for expressing it a binary detour phase computer generated hologram technique was used. For decryption, the correct key code, RPM has to be used. If the wrong key code is used then the decrypted image is noisy, which means

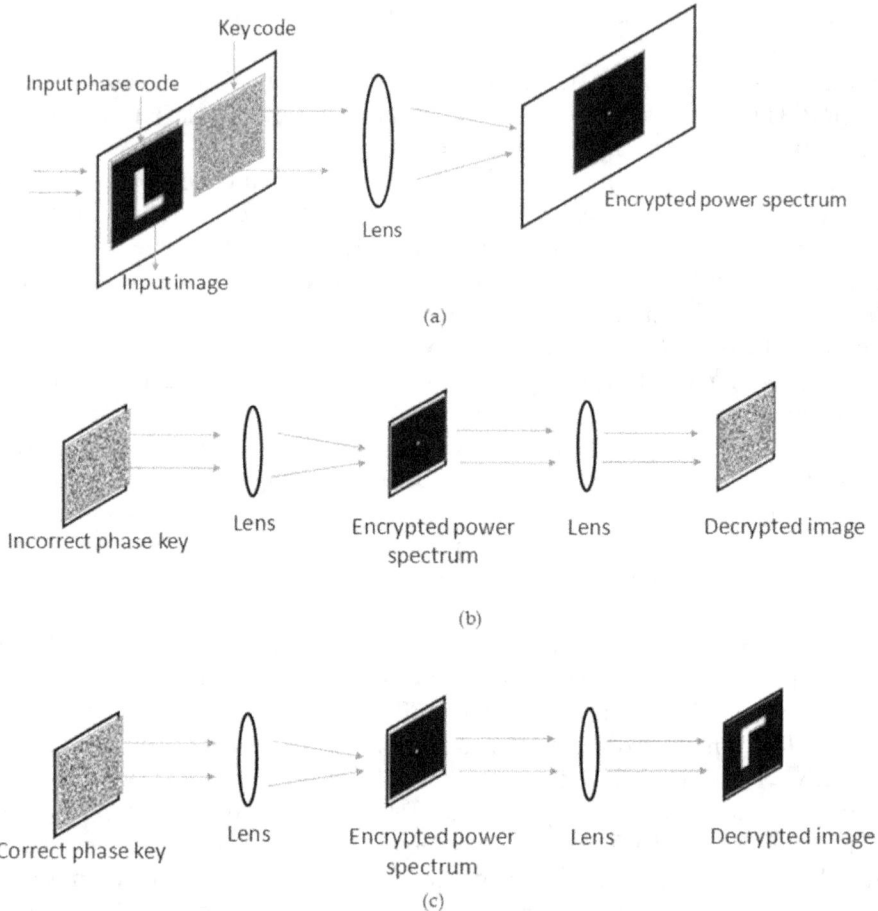

Figure 4.2. DRPE using JTC architecture; (a) encryption system, (b) decryption system with wrong RPM, and (c) decryption system with correct RPM.

that there is no retrieval of the original image, as shown in figure 4.2(b). With the use of the correct RPM, the original image is decrypted successfully, as shown in figure 4.2(c).

Results of computer simulation on a basic scheme of DRPE using JTC architecture are presented in figure 4.3. The real part of the input image, L and the key code, RPM in a single plane, which form the joint image, is shown in figure 4.3(a). The joint image is Fourier transformed and the encrypted JPS to be recorded by a CCD camera has been shown in figure 4.3(b). The real part of the key code, which has a complex value, to be fed into the decryption system is shown in figure 4.3(c). For expressing the complex valued RPM, a binary detour phase computer generated hologram technique can be used [7]. The decrypted image obtained with the correct and incorrect keys are shown in figures 4.3(d) and (e), respectively. It is clear from the results that, for successful retrieval of the original image, use of the correct RPM is a must.

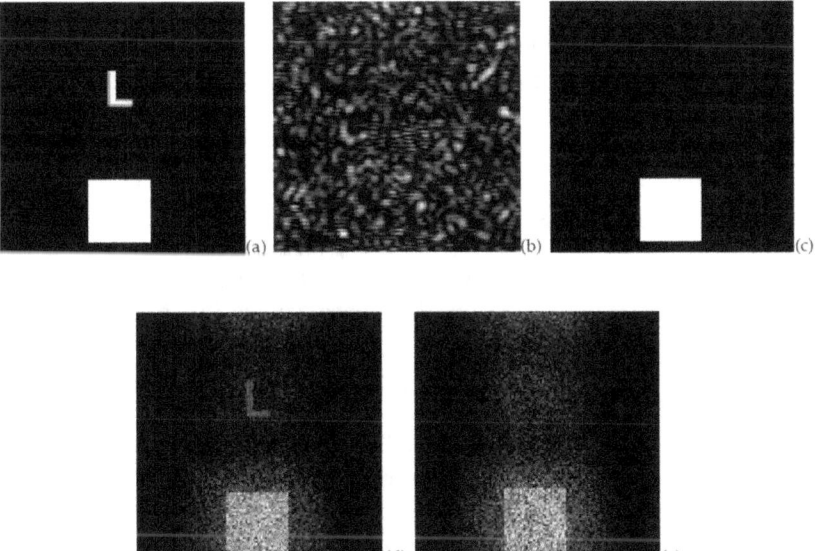

Figure 4.3. Simulation results. (a) The real part of the input image and the key code, RPM in a single plane, (b) encrypted joint power spectrum, (c) the real part of the key code to be fed into the decryption system, (d) the decrypted image with the correct key code, and (e) the decrypted image with the incorrect key code.

In order to improve the quality of the output images, a security system employing two numerically generated phase-only masks (POMs) were proposed [8]. The POMs were designed with an iterative POCS algorithm with constraints in the input and output domains. Further, an optical security system based on computer generated optical diffractive elements was reported, in which two phase-only transparencies were designed for a JTC. The resulting mask was used for security systems such that the desired code was received in the output plane only when the specific phase masks were placed in the input plane of JTC [9]. In a JTC architecture, the size of the joint input plane is doubled along one axis (horizontal or vertical) compared with the size of each input image, and the centre of each input image is shifted. In an improvement on removing the effect of phase terms in JTC and auto-correlation terms contributing the reconstruction of the original image, a JTC-based encryption system was demonstrated [10]. Using a photorefractive material, a secure optical storage system was also reported, in which the real-valued key code was designed through the use of an optimized algorithm [11]. For perfect reconstruction in a cryptosystem based on JTC, the key mask must be a phase-only function. However, this is not practical in a JTC scheme because inverse Fourier transform of the phase function should be embedded into a small window of the input plane. As a result, the truncated key mask no longer remains a phase-only function in the spectral domain. This aspect was studied in [12] and the key mask design was suggested.

The DRPE technique employing JTC architecture has undergone several modifications such as nonlinear JTC, FRT-based fractional JTC, and fully-phase JTC. Most of the variants of JTC have been combined with DRPE for image encryption and authentication applications. For the first time in 2006, a digital Fresnel hologram was recorded through three-step phase-shifting interferometry in JTC architecture [13]. The method has the benefit that the secured data can be transferred via digital communication channels and the encryption-decryption process can be achieved at high speed. The experiment was performed with a programmable liquid crystal television (LCTV) display working in pure phase mode. The LCTV was used for displaying both input signal and for introducing phase shift. Employing a nonlinear JTC in the Fresnel domain, an image encryption method has also been demonstrated [14]. The scheme used a lensless optical system and relatively simplified version of JTC. Based on the phase truncation approach and phase retrieval, an asymmetric scheme for authentication was reported, in which only one encrypted reference was required for the verification of multiple images [15].

In addition to an intensity image, phase-only functions have also been used for the study of authenticity in a conventional JTC scheme [16]. For a given specific target at the output plane, both the phase-only functions in the input plane can be generated using the phase retrieval algorithm. The second phase function is fixed and refers to as a lock and the other phase function is referred to as a key. Further, a multichanneled encryption by using multiple RPMs in JTC scheme has been reported [17]. Use of multiple wavelengths for color image encryption [18] and encryption through wavelength multiplexing [19] in JTC have also been reported.

Cryptanalysis of any cryptosystem is the most important study. When JTC-based cryptosystems started drawing the attention of scientific community, the attack analysis on those schemes were also reported. It was demonstrated that the JTC-based scheme showed vulnerability to known-plaintext and chosen-plaintext attacks [20, 21]. In a known-plaintext attack, the intruder knows a single plaintext–ciphertext pair in addition to the encryption method. In this case, the intruder can access keys in the signal and spatial frequency domains by using a phase retrieval algorithm. It is also possible that the key distributions can be found through heuristic attack, though the output will be noisy. In the chosen-plaintext attack, the intruder introduces an adequately designed input in the encryption-decryption system for determining the key distribution [22, 23].

The attack analysis of the JTC-based asymmetric cryptosystem has also been reported [24, 25]. Collision in the asymmetric cryptosystem based on phase-truncated Fresnel transform approach has been analyzed. The collision is defined as a phenomenon in which two distinct inputs produce identical output. If an attacker accesses the encryption keys in such a way, when it is applied to encrypted data it produces an arbitrary image instead of the original one [24]. In this study, for encryption, the conventional RPMs were replaced with structured phase masks (SPMs) with defined construction parameters. The decryption keys were created during amplitude- and phase-truncation process. An attacker generates an arbitrary (collision) image from the encrypted image using a modified Gerchberg–Saxton algorithm (MGSA) and the two different users; the authorized and unauthorized user (attacker), can claim the retrieved image as the original one. The authorized user uses the correct decryption keys and retrieves the original image, while the unauthorized user uses the generated keys and produces the collision image. The

JTC architecture verifies the authenticity of the retrieved data. An auto-correlation peak is obtained when image retrieved by the authorized user is matched with the encrypted image and a cross-correlation is obtained when the encrypted image is matched with the collision image.

Since the JTC-based cryptosystem was reported vulnerable to applicable attacks, its use in authentication got emphasis thereafter. The concept of photon counting imaging was introduced into the system for authentication applications [26–28]. In photon counting, images have a limited number of photons. This is achieved by controlling the expected number of incident photons. An input image was encrypted under the DRPE framework and a photon-limited encrypted image was obtained. The photon counting encrypted image was generated with few photons, which appeared sparse but was carrying sufficient information required for authentication. The sparse representation of the encrypted function produced a decoded image, which was not easily recognizable by intruders. Further, computational ghost imaging has also been reported for authentication application [29]. The reconstructed image was authenticated by using a significantly small number of realizations (i.e. less than 5% of Nyquist limit) rather than directly extracting the original object.

A method of image hiding was reported, which combined the JTC architecture with holography [30]. The hidden image encrypted through JTC was embedded into the Fourier hologram of a host image. The watermarked image was retrieved by obtaining the inverse Fourier transform and the hidden image was decoded from the watermarked hologram employing JTC. Alleviating the need of obtaining complex-valued encrypted image for decryption and verification, a method was proposed, which combined phase retrieval and median filtering [31]. In this case, sparse representation for the phase distribution was used for decryption and verification. Further, an authentication system based on compressed DRPE and quick response (QR) code was also proposed [32]. A cryptosystem employing JTC for securing 3-D objects has also been reported [33]. In this case, the encrypted object information is contained in the JPS.

4.3 Authentication using fractional non-conventional JTC

In an authentication system, the correlation peak detected by an intensity detector or a CCD camera is converted into an electronic signal for decision-making. If the signal strength is above some predefined threshold, the input key (RPM) verifies the true input. In this section, a fractional JTC with a non-conventional approach is explained [34].

A fractional correlator offers two main advantages over the conventional correlator; first, the width of the cross-correlation peak is narrower and secondly, the intensity and location of correlation peaks are related to the fractional orders [35]. Therefore, the desired intensity and position can be achieved by appropriately changing the fractional order. The ease in implementation and advantages in terms of performance measures over the conventional JTC scheme motivated the use of a fractional correlator in information security and authentication applications.

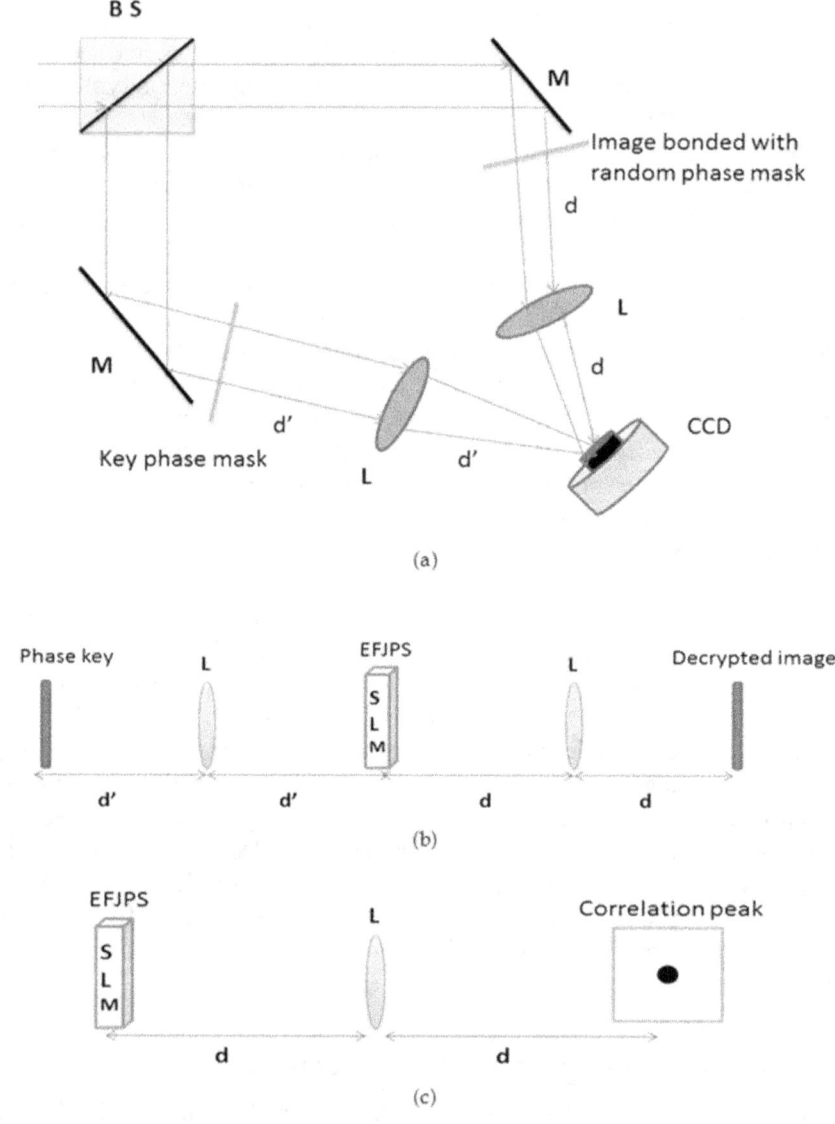

Figure 4.4. (a) Fractional non-conventional JTC-based authentication architecture: (a) encryption scheme, (b) decryption scheme, and (c) only for verification. L: lens, EFJPS: encrypted fractional joint power spectrum, CCD: charge-coupled device, SLM: spatial light modulator, BS: beam splitter, d & d': free space propagation distances for the order of FRT calculation.

An encryption and verification method based on nonconventional fractional JTC is shown in figure 4.4. For encryption, an input image bonded with an RPM and key phase mask are fractional Fourier transformed independently. The encrypted fractional joint power spectrum (EFJPS) can be recorded by an intensity sensing device. The encrypted power spectrum is referred to as the encrypted image. Figure 4.4(a) shows the optical scheme for encryption. Let $f(x,y)$ denote an image to be encrypted. The image $f(x,y)$ is multiplied with an RPM $\exp[i2\pi R_1(x,y)]$ is fractional Fourier transformed with some order α_1.

Let $g(\xi,\eta)$ represent the FRT of function $\{f(x,y) \times \exp[i2\pi R_1(x,y)]\}$ [9, 30]

$$g(\xi,\eta) = K \iint f(x,y) \times \exp[i2\pi R_1(x,y)]$$
$$\times \exp\left[i\pi\frac{x^2 + y^2 + \xi^2 + \eta^2}{\tan \alpha_1} - 2i\pi\frac{xy\xi\eta}{\sin \alpha_1}\right]dxdy \tag{4.3}$$

Here, K represents a complex constant. A key phase mask $\exp[i2\pi R_2(x,y)]$ is separately fractional transformed. Let $R(\xi,\eta)$ represent the FRT of order α_2 of this key phase mask.

$$R(\xi,\eta) = K \iint \exp[i2\pi R_2(x,y)]$$
$$\times \exp\left[i\pi\frac{x^2 + y^2 + \xi^2 + \eta^2}{\tan \alpha_2} - 2i\pi\frac{xy\xi\eta}{\sin \alpha_2}\right]dxdy \tag{4.4}$$

Functions $R_1(x,y)$ and $R_2(x,y)$ are independent random distributions with values between 0 and 1. The order of FRT is defined as $\alpha = p\pi/2$. Here 'p' is the order parameter of FRT whose value is chosen arbitrarily. The coordinates (x,y) and (ξ,η) represent the coordinates of spatial and fractional domains, respectively. Thus, the obtained spectra through equations (4.3) and (4.4) are combined together into what is called the EFJPS. This is also referred to as an encrypted image.

$$E(\xi,\eta) = g(\xi,\eta) \times R^*(\xi,\eta) \tag{4.5}$$

The symbol * denotes the complex conjugate. The schematic diagram for decryption is shown in figure 4.4(b). For decryption, the encrypted image, $E(\xi,\eta)$ is multiplied with the fractional spectrum of the key phase mask and the combined function is inverse fractional Fourier transformed with the same order. The output can be recorded by a CCD camera as a decrypted image, $d(x,y)$.

$$d(x,y) = \mathfrak{F}^{-\alpha_1}[E(\xi,\eta) \times R(\xi,\eta)]$$
$$= \mathfrak{F}^{-\alpha_1}[g(\xi,\eta)] \tag{4.6}$$
$$= f(x,y) \times \exp[j2\pi r_1(x,y)]$$

Here, $\mathfrak{F}^{-\alpha_1}$ represents the FRT operator with order—α_1. For obtaining the EFJPS, it is possible to use same orders of FRT. Use of different orders of FRT will provide additional keys to the cryptosystem. For optical implementation of this scheme, the EFJPS can be displayed over a suitable SLM. The EFJPS is a complex function, which can be separated into real and imaginary parts. It is possible that using either real or imaginary parts of the EFJPS, the original content can be retrieved successfully. Therefore, quantizing the EFJPS will help reduce the file size for transmission over communication channels.

4.3.1 Authentication

Figure 4.4(c) shows the schematic for authentication verification. This geometry along with the scheme shown in figure 4.4(a) is used for this purpose. The FRT of order α_1 of the original image bonded with RPM is given as

$$g(\xi,\eta) = K \iint f(x,y) \times \exp\left[i2\pi R(x,y)\right]$$
$$\times \exp\left[i\pi\frac{x^2 + y^2 + \xi^2 + \eta^2}{\tan\alpha_1} - 2i\pi\frac{xy\xi\eta}{\sin\alpha_1}\right]\mathrm{d}x\mathrm{d}y \tag{4.7}$$

The FRT of order α_2 of the key phase mask (identical to the RPM bonded with input image) is given as

$$R_1(\xi,\eta) = K \iint \exp[i2\pi R(x,y)]$$
$$\times \exp\left[i\pi\frac{x^2 + y^2 + \xi^2 + \eta^2}{\tan\alpha_2} - 2i\pi\frac{xy\xi\eta}{\sin\alpha_2}\right]\mathrm{d}x\mathrm{d}y \tag{4.8}$$

The power spectrum is given as

$$E'(\xi,\eta) = g(\xi,\eta) \times R_1^*(\xi,\eta) \tag{4.9}$$

Computing fractional transform of the EFJPS, $E'(\xi,\eta)$ for some order α_3 would produce a sharp auto-correlation peak, if the same RPMs are used during encryption [34, 35].

$$C(u,v) = K \iint g(\xi,\eta)R_1^*(\xi,\eta)$$
$$\times \exp\left[i\pi\frac{u^2 + v^2 + \xi^2 + \eta^2}{\tan\alpha_3} - 2i\pi\frac{uv\xi\eta}{\sin\alpha_3}\right]\mathrm{d}\xi\mathrm{d}\eta \tag{4.10}$$

For simplifying the expressions, two new functions can be introduced, as

$$f_1(x,y) = f(x,y) \times \exp[i2\pi R(x,y)] \tag{4.11}$$

$$f_2(x_1,y_1) = \exp[i2\pi R(x_1,y_1)] \tag{4.12}$$

Equation (4.10) can now be written as

$$C(u,v) = K \iint \iint \{f_1(x,y) \times f_2^*(x_1,y_1)\}$$
$$\times \exp\left[i\pi\frac{x^2 + y^2 + \xi^2 + \eta^2}{\tan\alpha_1} - i\pi\frac{x_1^2 + y_1^2 + \xi^2 + \eta^2}{\tan\alpha_2} + i\pi\frac{u^2 + v^2 + \xi^2 + \eta^2}{\tan\alpha_3}\right] \tag{4.13}$$
$$\times \exp\left[-2i\pi\frac{xy\xi\eta}{\sin\alpha_1} + 2i\pi\frac{x_1y_1\xi\eta}{\sin\alpha_2} - 2i\pi\frac{\xi\eta uv}{\sin\alpha_3}\right]\mathrm{d}x\mathrm{d}y\mathrm{d}x_1\mathrm{d}y_1\mathrm{d}\xi\mathrm{d}\eta$$

If the following condition

$$\frac{1}{\tan \alpha_1} - \frac{1}{\tan \alpha_2} + \frac{1}{\tan \alpha_3} = 0 \tag{4.14}$$

is satisfied, the quadratic phase terms with coordinates ξ and η in equation (4.13) are eliminated. Then the integral becomes simply a Dirac delta function, $\delta(.)$, which has a value of one when its argument is equal to zero, or zero otherwise.

$$\iint \exp\left[-2i\pi\frac{xy\xi\eta}{\sin \alpha_1} + 2i\pi\frac{x_1y_1\xi\eta}{\sin \alpha_2} - 2i\pi\frac{\xi\eta uv}{\sin \alpha_3}\right]d\xi d\eta$$

$$= \delta\left(-\frac{x}{\sin^2 \alpha_1} + \frac{x_1}{\sin^2 \alpha_2} - \frac{u}{\sin^2 \alpha_3}\right)\delta\left(-\frac{y}{\sin^2 \alpha_1} + \frac{y_1}{\sin^2 \alpha_2} - \frac{v}{\sin^2 \alpha_3}\right) \tag{4.15}$$

From the zero argument of equation (4.15),

$$x_1 = \left(\frac{x}{\sin^2 \alpha_1} + \frac{u}{\sin^2 \alpha_3}\right)\sin^2 \alpha_2 \tag{4.16}$$

$$y_1 = \left(\frac{y}{\sin^2 \alpha_1} + \frac{v}{\sin^2 \alpha_3}\right)\sin^2 \alpha_2 \tag{4.17}$$

considering simple choices for the orders (α_1, α_2, α_3) to satisfy equation (4.14) that represent the correlation of $f_1(x,y)$ and $f_2(x_1,y_1)$. Assuming, $\alpha_1 = \alpha_2$ and $\alpha_3 = \pi/2$, then equations (4.16) and (4.17) become,

$$x_1 = x + u \sin^2 \alpha_2 \tag{4.18}$$

$$y_1 = y + v \sin^2 \alpha_2 \tag{4.19}$$

The intensity distribution of joint fractional correlation is given by

$$C(u,v) = K\iint \{f_1(x,y) \times f_2^*(x + u \sin^2 \alpha_1, y + v \sin^2 \alpha_1)\}$$

$$\times \exp\left[-i\pi\frac{xu \sin^2 \alpha_1}{\tan \alpha_1} - i\pi\frac{yv \sin^2 \alpha_1}{\tan \alpha_1}\right]dxdy \tag{4.20}$$

At the centre ($u = 0$ and $v = 0$) of the correlation output plane, the value is given as

$$C(0,0) = K\iint \exp[i2\pi R(x,y)] \times f(x,y)\exp[i2\pi R(x,y)]dxdy \tag{4.21}$$

The mathematical analysis explains that to have an auto-correlation peak, there are two important pieces of information: the same RPMs must be used and FRTs of the same orders calculated. If the orders of FRT used during the verification process are not identical then cross-correlations would appear (figure 4.5).

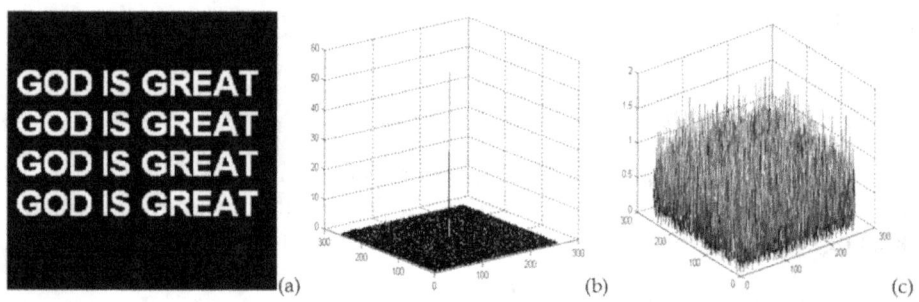

Figure 4.5. Simulation results. (a) Binary text image to be encrypted, (b) autocorrelation peak obtained after using all correct keys, and (c) the cross-correlation peak obtained after using the wrong RPM. Reprinted from [34], Copyright (2012), with permission from Elsevier.

4.3.2 Authentication with a phase-encoded image

Chapter 3 discusses the fully-phase encryption-decryption procedure. It has been stated that a fully-phase image offers enhanced security as compared to the intensity image. Because of noise tolerant properties, it provides better discrimination capability. In this section, the image authentication verification for a phase-encoded image is discussed. For this study, an input image is first phase-encoded and then it is multiplied with an RPM whose fractional transform is calculated. All the steps discussed in the previous section for the verification process are repeated for a phase-encoded function [34].

4.3.3 Performance measurement

Any cryptosystem cannot be claimed secure unless it is able to endure various attacks. Therefore, cryptanalysis of the encryption-decryption scheme is necessary. The parameters such as mean square error (MSE), image mean square signal-to-noise ratio (SNR), and peak signal-to-noise ratio (PSNR) are important to analyze the cryptosystems. Similar to the performance measurement of the cryptosystem, it is important to study the performance measure parameters for authentication applications. The study of various parameters to analyze the correlation output is discussed in this section.

The MSE between the original and decrypted image at the pixel (i,j) is defined as:

$$\begin{aligned}
\text{MSE} &= \| I(x,y) - D(x,y)\|^2 \\
&= \frac{1}{N \times N} \sum_{i=1}^{N} \sum_{j=1}^{N} |I(x,y) - \text{abs}[D(x,y)]|^2
\end{aligned} \tag{4.22}$$

where $I(x,y)$ and $D(x,y)$ represent the input and decrypted images, respectively. Here, $N \times N$ represents the size of the input image.

The image mean square SNR between the original and decrypted image is defined as

$$\text{Mean square SNR} = \frac{\sum_{x=1}^{N}\sum_{y=1}^{N}[I(x,y)]^2}{\sum_{x=1}^{N}\sum_{y=1}^{N}[I(x,y) - D(x,y)]^2} \tag{4.23}$$

Higher mean square SNR values indicate better noise tolerance.

For an image of size $N \times N$ pixels, the PSNR is defined as

$$\text{PSNR} = 10 \log_{10}\frac{(N-1)^2}{\frac{1}{N \times N}\sum_{i=0}^{N-1}\sum_{j=0}^{N-1}(I(x,y) - D(x,y))^2} \tag{4.24}$$

To indicate the degree of transparency, the PSNR is calculated between the original and decrypted images.

The performance of an authentication verification systems are computed through various parameters such as peak intensity (PI), discrimination ratio (DR), SNR, peak-to-sidelobe-ratio (PSR), peak-to-correlation energy (PCE) [34].

The discrimination ratio (DR) is defined as

$$\text{DR} = 1 - \frac{\text{Highest peak due to crosscorrelation}}{\text{Autocorrelation peak height}} \tag{4.25}$$

The SNR measures the sensitivity of the auto-correlation peak to the additive noise at the input plane. Mathematically it is defined by

$$\text{SNR} = \frac{[E|C_{I,I+n}(0,0)|]^2}{VAR|C_{I,I+n}(0,0)|} \tag{4.26}$$

where I is the input image, n is an additive noise, and C is the correlation between input and output signal. Here $E\{.\}$ represents the expected value and VAR stands for the variance. The variance arises due to noise.

The PSR measures the sharpness of the auto-correlation peak. The maximum output correlation value is defined by

$$R_0^2 = \max[|C(u,v)|^2] \tag{4.27}$$

where R_0^2 is the output correlation-peak value and $C(u,v)$ is the correlation output value at (u,v). The maximum sidelobe (SL) intensity in the correlation output plane based on R_0^2, outside the annular region is expressed as

$$\text{SL} = \max_{\substack{u,v > \text{circular} \\ \text{region with} \\ \text{diameter } R_0^2/2}} [|C(u,v)|^2] \tag{4.28}$$

The PSR is a measure of peak sharpness. It is defined as

$$\text{PSR} = \frac{R_0^2}{SL} \tag{4.29}$$

In terms of mean, μ and standard deviation σ of correlation coefficient in the region around peak intensity, the PSR is defined as

$$\text{PSR} = \frac{\text{Peak} - \mu}{\sigma} \qquad (4.30)$$

The PCE is defined as

$$\text{PCE} = \frac{|c(0,0)|^2}{E_y} \qquad (4.31)$$

where $c(0,0)$ denotes the correlation peak intensity. E_y represents the correlation plane energy, which is defined as

$$E_y = \int\int_{-\infty}^{\infty} c(x,y)\mathrm{d}x\mathrm{d}y \qquad (4.32)$$

The values of PI, SNR, and PSR are calculated for the input image. In the authentication study, the PSR characterizes how well the key phase mask (RPM) matches with the phase mask bonded with the input image. If the values of these parameters are computed for an intensity image and fully-phase image, then it is found that the phase-encoded image has much larger values as compared to the intensity image [34]. This suggests the use of a phase-encoded image for authentication applications as compared to an intensity image. Recently, a hash-based authentication system employing fractional JTC and equal modulus decomposition has been proposed [36]. Hash functions are used for obtaining the digital signature of data, which is a compressed form of the input data. Most recently, an authentication scheme using Stokes polarimetry of vector beams have also been reported [37]. In this method, phase-only functions of the plaintext have been generated using a modified GS algorithm, which are used to tailor the phase of a vector beam. Applying the principle of multiple-input JTC, multiple secured images/ data can be authenticated simultaneously [38]. To further improve the performance of the correlation scheme for verification applications, a nonlinear correlator can be combined.

MATLAB code

```
%JTC based encryption
%The Matlab code uses the JTC architecture to perform optical encryption
%%%%%%%%%%%%%%%%%%%%% %%%%%%%%%%%%%%%%%%%%% %%%%%%%%%%%%%%%%%%%%%
  %%%%%%%%%%%%%%%%%%
N = 128;
```

```
A = zeros(N);

a1 = imread ('L_32.bmp');  %Read the input image to be encrypted
a1 = double(a1(:,:,1));
a1 = a1./max(max(a1));
figure;imagesc(abs(a1));colormap(gray);


key1=exp(2*pi*1i.*rand(32));  %Random phase mask (encryption key)
  to be bonded with input image
a1=a1.*key1;

key2=exp(2*pi*1i.*rand(32));  %Random phase key code

%Placing the phase bonded input image and the key code in one plane
A(28:59,48:79) = a1;
A(91:122,48:79) =key2;

figure,imagesc(abs(A));colormap(gray);
title('Input for encryption system');

ft_A = fftshift(fft2(ifftshift(A)));
jps = ft_A.*conj(ft_A);  %Joint power spectrum
figure;imagesc(abs(jps));colormap(gray);

%Decryption process
C=zeros(32);
B=zeros(N);
B(28:59,48:79) = C;
B(91:122,48:79) = key2;  %Phase key code;
figure;imagesc(abs(B));colormap(gray);
title('Input for decryption system');

corr1 = fftshift(fft2(ifftshift(B)));
corr2=corr1.*jps;  %Multiplication with joint power spectrum
decrypted=fftshift(ifft2(ifftshift(corr2)));
figure;imagesc(abs(decrypted));colormap(gray);
%Test decryption with incorrect key
key2_incorrect=exp(2*pi*1i.*rand(32));  %incorrect key for test

B(28:59,48:79) = C;
B(91:122,48:79) = key2_incorrect;  %incorrect phase key code;
figure;imagesc(abs(B));colormap(gray);
title('Input for decryption system');
corr1 = fftshift(fft2(ifftshift(B)));
corr2=corr1.*jps;  %Multiplication with joint power spectrum
```

```
decrypted_incorrect=fftshift(ifft2(ifftshift(corr2)));
figure;imagesc(abs(decrypted_incorrect));colormap(gray);
```

References

[1] VanderLugt A B 1964 Signal detection by complex filters *IEEE Trans. Inf. Theory* **10** 139–45
[2] Weaver C S and Goodman J W 1966 Techniques for optically convolving two functions *Appl. Opt.* **5** 1248–9
[3] Javidi B and Horner J L 1994 Optical pattern recognition for validation and security verification *Opt. Eng.* **33** 1752–6
[4] Javidi B and Wang J 1996 Position-invariant two-dimensional image correlation using a one-dimensional space integrating optical processor: application to security verification *Opt. Eng.* **35** 2479–86
[5] Brasher J D and Johnson E G 1997 Incoherent optical correlators and phase encoding of identification codes for access control or authentication *Opt. Eng.* **36** 2409–16
[6] Weber D and Trolinger J 1999 Novel implementation of nonlinear joint transform correlators in optical security and validation *Opt. Eng.* **38** 62–8
[7] Nomura T and Javidi B 2000 Optical encryption using joint transform correlator architecture *Opt. Eng.* **39** 2031–5
[8] Li Y, Kreske K and Rosen J 2000 Security and encryption optical systems based on a correlator with significant output images *Appl. Opt.* **39** 5295–301
[9] Abookasis D, Arazi O, Rosen J and Javidi B 2001 Security optical systems based on joint transform correlator with significant output images *Opt. Eng.* **40** 1584–9
[10] Park S J, Kim J Y, Bae J K and Kim S J 2001 Fourier plane encryption technique based on removing the effect of phase terms in a joint transform correlator *Opt. Rev.* **18** 413–5
[11] Nomura T, Mikan S, Morimoto Y and Javidi B 2003 Secure optical data storage with random phase key codes by use of a configuration of a joint transform correlator *Appl. Opt.* **42** 1508–14
[12] Lin L C and Cheng C J 2005 Optimal key mask designs for optical encryption based on joint transform correlator architecture *Opt. Commun.* **285** 144–54
[13] Mela C L and Iemmi C 2006 Optical encryption using phase-shifting interferometry in a joint transform correlator *Opt. Lett.* **31** 2562–4
[14] Vilardy J M, Millán M S and Pérez-Cabré E 2014 Nonlinear optical security system based on a joint transform correlator in the Fresnel domain *Appl. Opt.* **53** 1674–82
[15] Rajput S K and Nishchal N K 2014 An optical encryption and authentication scheme using asymmetric keys *J. Opt. Soc. Am.* A **31** 1233–8
[16] Chang H T and Chen C C 2006 Fully phase asymmetric image verification system based on joint transform correlator *Opt. Express* **14** 1458–67
[17] Amaya D, Tebaldi M, Torroba R and Bolognini N 2008 Multichanneled encryption via a joint transform correlator architecture *Appl. Opt.* **47** 5903–7
[18] Amaya D, Tebaldi M, Torroba R and Bolognini N 2008 Digital color encryption using a multiwavelength approach and a joint transform correlator *J. Opt. A: Pure Appl. Opt.* **10** 104031
[19] Amaya D, Tebaldi M, Torroba R and Bolognini N 2009 Wavelength multiplexing encryption using joint transform correlator architecture *Appl. Opt.* **48** 2099–104

[20] Barrera J F, Vargas C, Tebaldi M, Torroba R and Bologini N 2010 Known plaintext attack on a joint transform correlator encrypting system *Opt. Lett.* **35** 3553–5

[21] Barrera J F, Vargas C, Tebaldi M and Torroba R 2010 Chosen plaintext attack on a joint transform correlator encrypting system *Opt. Commun.* **283** 3917–21

[22] Qin W, Peng X and Meng X 2011 Cryptanalysis of optical encryption schemes based on joint transform correlator architecture *Opt. Eng.* **50** 028201

[23] Lin C and Shen X J 2012 Analysis and design of impulse attack free generalized joint transform correlator optical encryption scheme *Opt. Laser Technol.* **44** 2032–6

[24] Mehra I, Rajput S K and Nishchal N K 2013 Collision in Fresnel domain asymmetric cryptosystem using phase truncation and authentication verification *Opt. Eng.* **52** 028202

[25] Mehra I, Rajput S K and Nishchal N K 2014 Cryptanalysis of an image encryption scheme based on joint transform correlator with amplitude- and phase-truncation approach *Opt. Lasers Eng.* **52** 167–73

[26] Perez-Cabre E, Cho M and Javidi B 2011 Information authentication using photon-counting double-random-phase encrypted images *Opt. Lett.* **36** 22–4

[27] Rajput S K, Kumar D and Nishchal N K 2015 Optical encryption system based on phase mask multiplexing and photon counting imaging for multiple image authentication and digital hologram security *Appl. Opt.* **54** 1657–66

[28] Rajput S K, Kumar D and Nishchal N K 2014 Photon counting imaging and polarized light encoding for secure image verification and hologram watermarking *J. Opt.* **16** 125406

[29] Chen W and Chen X 2013 Object authentication in computational ghost imaging with the realizations less than 5% of Nyquist limit *Opt. Lett.* **38** 546–8

[30] Shi X and Zhao D 2011 Image hiding in Fourier domain by use of joint transform correlator architecture and holographic technique *Appl. Opt.* **50** 766–72

[31] Wang X, Chen W and Chen X 2015 Optical encryption and authentication based on phase retrieval and sparsity constraints *IEEE Photon. J* **7** 7800310

[32] Wang X, Chen W and Chen X 2015 Optical information authentication using compressed double-random-phase-encoded images and quick-response codes *Opt. Express* **23** 6239–53

[33] Zea A V, Ramirez J F B and Torroba R 2016 Three-dimensional joint transform correlator cryptosystem *Opt. Lett.* **41** 599–602

[34] Rajput S K and Nishchal N K 2012 Image encryption and authentication verification scheme using fractional nonconventional joint transform correlator *Opt. Lasers Eng.* **50** 1474–83

[35] Lohmann A W and Mendlovic D 1997 Fractional joint transform correlator *Appl. Opt.* **36** 7402–7

[36] Kumar A, Fatima A and Nishchal N K 2018 An optical Hash function construction based on equal modulus decomposition for authentication verification *Opt. Commun.* **428** 07–14

[37] Fatima A and Nishchal N K 2018 Image authentication using vector beam with sparse phase information *J. Opt. Soc. Am.* A **35** 1053–62

[38] Nishchal N K, Goyal S, Aran A, Beri V K and Gupta A K 2006 Binary differential joint transform correlator based on a ferroelectric liquid crystal electrically addressed spatial light modulator *Opt. Eng.* **45** 026401 10

Chapter 5

Image watermarking and hiding

5.1 Introduction

In this digital era, the problem of unauthorized access and dissemination of information has become a serious issue due to the rapid development of the Internet and modern communication gadgets. A digital or optical cryptosystem permits access to the secured data only to valid key holders (authentic users). However, once such data is decrypted with the application of the correct keys, it becomes almost infeasible to track its reproduction. The digital data can be duplicated easily without any loss of quality, due to the availability of modern copiers, scanners and printers. Therefore, digital products attract hackers' attention, which may allow unauthorized and illegal access and use of information. This is called *data piracy*. The piracy of information deprives the rights of original creators and harms innovation.

A digital watermark (discrete data stream) is a visible or invisible identification mark that is embedded into the data and remains present after the decryption process. The watermarking is a potential and effective way to protect the copyright and encryption-decryption problems. The major driving force for research in watermarking is the requirement for effective copyright protection for digital images, audio, video, and text. This has become an important issue due to the rapid growth of the Internet and availability of multimedia computing facilities. The information content in the digital format can be used without taking permission of the original owner. In these applications, some pattern or number is watermarked into the signal to protect the ownership. The schematic of the digital watermarking process has been shown in figure 5.1. The encoder combines the information to be embedded with original data, which is known as marked data. For retrieving the information, received data is processed by the decoder.

Historically, watermarking is connected with the steganography technique. Steganography is defined as the science of writing covered messages, which cannot be detected easily. The information is hidden in such a way that it has no relation

doi:10.1088/978-0-7503-2220-1ch5

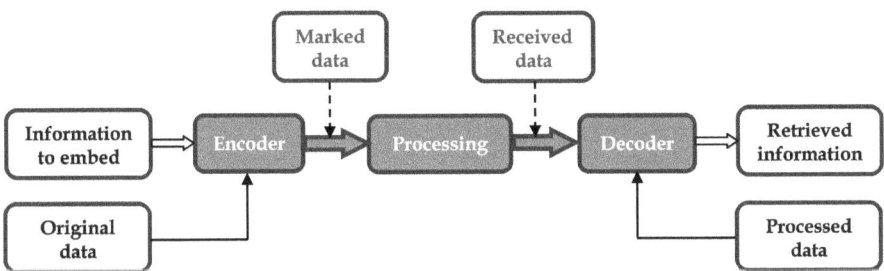

Figure 5.1. Watermarking process.

with the host. It is created with the aim that only the sender and intended recipient can share the message. To any other person, the access of information is prevented. The watermarking process may not necessarily hide the fact of secret transmission of message but it allows both the parties to communicate secretly. There are several applications where watermarking can be used, such as broadcast monitoring, copy control, authentication, transaction tracking, and copyright protection. It is expected that an attacker will always try to attempt to remove the watermark by modifying the watermarked message/data. The watermark must survive the common signal processing operations and counterfeit attempts. Therefore, it is always strived to embed the watermark in such a way that it is difficult to remove, without applying the correct keys, unless the watermarked signal is distorted. For an effective watermarking and detection process, the important properties are:

(i) **Unobtrusiveness**: it means the watermark should be perceptually invisible. There should be no interference between the watermark and host.

(ii) **Robustness**: the removal of the watermark should be difficult and the watermark should be immune to common signal processing techniques and geometric distortions.

(iii) **Universality**: the watermark should be applicable to the image, audio, video, text, and data.

(iv) **Unambiguous**: the watermark should be unique and retrievable.

The watermarking is considered to complement cryptography. A good watermarking method should satisfy a number of conditions. For example, due to watermarking the quality of host image should not be affected significantly and there should be no compromise with the level of security. In other words, decoding the hidden information by any unintended recipient should be hard enough, even if the watermarking scheme is known. Also, a watermark should not be lost during common image processing operations such as cropping, enlarging or reducing the watermarked content. The fundamental issues with the information hiding and watermarking are fidelity, robustness, payload, and security. *Fidelity* is defined as the degree of perceptual degradation due to embedding. *Robustness* is defined as the immunity against all forms of manipulation, which could be intentional or non-intentional. *Payload* is defined as the amount of message that can be embedded and extracted reliably.

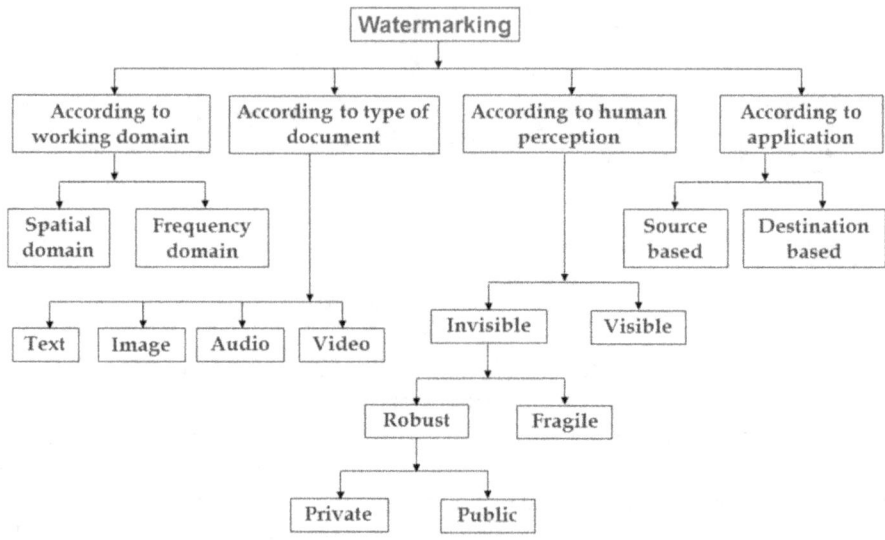

Figure 5.2. Types of watermarks.

Watermarks are categorized into various types depending on working domain, type of document, human perception, and application. Figure 5.2 depicts the classification tree of watermarks. A transformation domain is often needed for embedding the watermark robustly. Therefore, according to the working domain, the watermarks are categorized into object domain and frequency domain. Watermarking in the object domain is easier to implement and in this case, for watermark detection the original image/data is not required. However, such a scheme usually fails under common image processing attacks. In addition, fidelity of the original content can be degraded severely because the watermark is embedded directly on the pixel values. On the other hand, frequency domain watermarking resists most of the signal processing attacks and therefore provides higher protection. The technique depends on the spread spectrum approach and, therefore, the watermark becomes undetectable. It has been shown that embedding a watermark in the frequency domain is more robust than in the object domain [1–4].

According to the types of documents, the watermarks are categorized as text watermark, image watermark, audio watermark, and video watermark. Depending upon human perception, the watermarks are classified as invisible and visible watermarks. Invisible watermarks are usually preferred, which are further classified into robust and fragile watermarks. Depending upon the application, there are source-based and destination-based watermarks.

Another important issue is the secure communication amongst the parties. In order to conceal the existence of communications, information hiding and steganography are followed. Steganography is used for covert communication and can take increased payload size but with compromise in robustness. It has been combined with image encryption for enhancing the level of security.

The wavelet transform is a mathematical tool, which finds application in several fields. Compared with the Fourier transform, the WT has an advantage in achieving spatial and frequency localization and multi-resolution analysis features. In digital signal and image processing, the DWT is closely related to filter banks. DWT is the most widely adopted transformation technique for digital watermarking, which are applied in the frequency domain. The use of wavelet has been reported in blind watermark detection, in which the embedded watermark is retrieved without any information from the original content. Identifying the wavelet coefficients of the embedded watermark in the frequency domain for watermark detection is a difficult task [2]. The sensitivities of the human visual system are different for different frequencies. The detection of low frequency content is easier. Therefore, a wavelet-based watermarking algorithm provides both perceptual invisibility and robustness to compression because the embedding is done in the middle-frequency range [3].

5.2 Information hiding and watermarking under the DRPE framework

Of late, optical techniques for encryption and watermarking have generated a considerable amount of interest. It may be due to the scalability, size of the encryption-decryption keys, processing speed as compared to their purely digital counterparts. Therefore, the classical DRPE method has been applied for encryption, hiding, and watermarking [5–18]. The concept of digital holography has been combined with watermarking. In this technique, a randomly phase-modulated watermark and its Fourier frequency were superposed on an image. For recovering the watermark, a numerical reconstruction algorithm was applied [5].

Image/data hiding is a method for protecting information from unauthorized distribution or malicious use. In the image hiding technique, an image to be hidden is inserted into another image, called the host image. The host image should not suffer from any degradation due to this embedding process. Depending on the domain at which the image/data is inserted, the methods of data hiding are classified into the spatial domain and Fourier domain hiding methods. Hiding data in the spatial domain is easier to implement but suffers from degradation in the host image quality and it is not robust for the signal processing operations. On the other hand, Fourier domain hiding is robust for signal processing operations as compared to spatial domain hiding. Spatial domain information hiding based on the DRPE scheme has been reported, in which the problem of spatial domain embedding was controlled through the amplitude of the embedded image that suppressed the undesirable effects on the host image [6].

The encrypted image obtained in a DRPE scheme as discussed in chapter 2, can be expressed as

$$E(x,y) = \iint [E_1(u,v) \times \exp\{i2\pi R_2(u,v)\}] \exp[2\pi i(ux + vy)]dudv \qquad (5.1)$$

For embedding the watermark, a host image is required. The encrypted image is embedded into the host image. Denoting the host image by $H(x,y)$, the watermarked image or the transmitted image, $W(x,y)$ is given as

$$W(x,y) = aE(x,y) + H(x,y) \qquad (5.2)$$

where a is an arbitrary constant, called the weighting factor of the watermark. Its value is chosen so as to ensure the invisibility and robustness of the hidden image against distortions. Since the encrypted image is a complex function, the water-marked image $W(x,y)$ is also a complex function. For hidden image retrieval, DRPE is followed in reverse process.

In 2003, an optical method for information watermarking of 3D objects was reported [7]. Following the DRPE scheme in a phase-shift digital hologram, a hidden image was embedded into a 3D object. The watermarked hologram was decoded to reconstruct the hidden image and the 3D object. It was also shown that the hidden image could be recovered from a windowed watermarked hologram in order to increase the processing speed and reconstruct the 3D object from a different viewing angle. Further, a method to watermark a 3D object with another hidden 3D object using digital holography was presented [8]. In this study, it was claimed that the watermarked hologram is very secure because of the multi-key nature of the watermarking process. In all the examples of watermarking, the basic principle as expressed in equation (5.2) is followed.

The encryption and watermark embedding domains have been extended from the Fourier to Fresnel, fractional Fourier, gyrator and discrete cosine transform domains with the idea of enhanced key size and better security. A method for multiple image encryption and watermarking by random phase matching, which can encrypt and decrypt more than one image with the same set of transmitted patterns based on the idea of DRPE and the wave field superposition, was also reported [9]. An electro-optical method employing a JTC architecture for deciphering a water-mark from a concealogram has been demonstrated [10]. In this technique, one can conceal an image within another image by manipulating the halftone coding of the host image.

Reconstruction of an image from its diffraction field requires both the amplitude and phase information of the field. In a study, it was shown that the phase information yields a better result of image retrieval than the real or imaginary part. The recovered image from the phase information was found satisfactory especially for binary inputs. This method was applied for image encryption and watermarking, which can reduce the communication load for Internet transmission of secured data [11]. In a further study, an iterative Fourier transform algorithm was combined with DRPE for digital watermarking. The algorithm was used to generate two binary phase structures, in which one was used to encode the hidden image and the other as the encryption key [12]. The image watermarking method with the use of digital holography and embedding in the discrete-cosine transform domain was also reported [13]. It was claimed that this method could reduce the degradation on the superposed image. Image watermarking in the gyrator domain employing phase retrieval algorithm and chaos were also reported [14, 15]. Robustness of the DRPE spread-space spread-spectrum watermarking scheme was studied in detail in [17]. A watermarking scheme for light field imaging operating on its 4D Fourier spectrum has also been reported recently [18]. The scheme showed no significant impact on the quality of the refocused images or the depth map. Deriving the concept of the devil's staircase function, a devil's vortex Fresnel lens has been designed to be used for phase image watermarking. The devil's vortex lens is basically a phase-only devil's lens modulated by a helical structure [19].

A technique for optical security based on photon counting imaging and polarized light encoding for authentication and watermarking has been reported [20]. An input image was encoded using the concept of polarized light, which is parameterized using the Stokes–Mueller formalism. The scheme was used for securing 3D information through digital hologram watermarking. The block diagram of photon counting imaging and polarized light encoding-based digital hologram watermarking and authentication scheme is shown in figure 5.3. In the photon counting imaging technique, images have a limited number of photons with respect to the expected number of incident photons. Hence, a photon-limited version of the encrypted image is generated and as a consequence, a sparse representation of the encrypted image is used for decryption and authentication. These types of encrypted images give a decrypted image, which cannot be easily seen by the attackers. A detailed discussion on photon counting imaging can be found in chapter 10.

5.2.1 FRT domain watermarking

In FRT domain watermarking under the DRPE framework, the encrypted image is watermarked into a host image. Watermark embedding in the FRT domain has an additional advantage over embedding in a spatial or in frequency domain. The FRT offers a continuum of planes in which watermarks can be embedded. This approach uses the combination of object and frequency domains for embedding, which provides additional security against attackers. The fractional orders provide extra degrees of freedom to enlarge the key space [21–24]. The benefits apply to other transforms as well such as Fresnel, gyrator, and fractional random transforms.

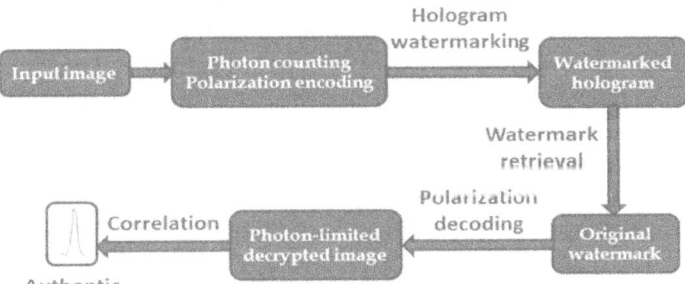

Figure 5.3. Block diagram of photon counting imaging and polarized light encoding based digital hologram watermarking and authentication scheme.

A basic watermarking scheme based on FRT domain encoding is discussed as an example. Let function $f(x,y)$ represent a 2D watermark to be encrypted in the FRT domain under the DRPE scheme. The watermark is bonded with an RPM, exp $\{i2\pi R_1(x,y)\}$ and its FRT for some order α_1 ($\alpha_1 = p_1\pi/2$) is obtained.

$$g(u,v) = K \iint f(x,y) \times \exp[2\pi i R_1(x,y)]$$
$$\times \exp\left(j\pi \frac{x^2 + y^2 + u^2 + v^2}{\tan \alpha_1} - 2j\pi \frac{xyuv}{\sin \alpha_1}\right) \mathrm{d}x\mathrm{d}y \qquad (5.3)$$

Here, (x,y) and (u,v) denote the coordinates of the object and FRT plane, respectively, and K is a complex constant, which has been defined in equation (2.22). The second step is to bond the obtained spectrum with a statistically independent RPM, $\exp\{i2\pi R_2(u,v)\}$ and get the resultant again fractional Fourier transformed for order α_2 ($\alpha_2 = p_2\pi/2$). The value for p_1 and p_2 are taken between 0 and 1. The obtained expression is given as

$$E(\xi,\eta) = K \iint \{g(u,v) \times \exp[2\pi i R_2(u,v)]\}$$
$$\times \exp\left(i\pi \frac{\xi^2 + \eta^2 + u^2 + v^2}{\tan \alpha_2} - 2i\pi \frac{\xi\eta uv}{\sin \alpha_2}\right) \mathrm{d}\xi\mathrm{d}\eta \qquad (5.4)$$

The function $E(\xi,\eta)$ is the encrypted image of watermark $f(x,y)$. The encrypted version of watermark $E(\xi,\eta)$ is combined with a host image, $H(\xi,\eta)$. Thus, the watermarked image $W(\xi,\eta)$ is given by

$$W(\xi,\eta) = H(\xi,\eta) + aE(\xi,\eta) \qquad (5.5)$$

The arbitrary constant a ensures invisibility of the watermarked image. Usually, the value of a is chosen by trial and error. The watermark embedding into the host image should be performed in such a way that it is not perceptible to the human eye but it must be robust against the attacks. Also, the host image should not be degraded much. Therefore, the search of an optimum-weighting factor is crucial to the watermarking technique. In a study, it was shown that the optimal weighting factor a produces the least errors in the reconstructed 3D host object and the decoded watermark even in the presence of an occlusion attack [22]. This was based on a formula derived for the optimum-weighting factor with which the watermark was embedded into the digital hologram of a 3D host image. The value minimized the total MSE of the reconstructed 3D host object and the decoded watermark. The results of computer simulation as an example have been shown in figure 5.4.

A gray-scale host image has been shown in figure 5.4(a) and a binary text used as the watermark is shown in figure 5.4(b). The encrypted image of the watermark obtained with FRT orders $\alpha_1 = 0.25$, $\alpha_2 = 0.55$ is shown in figure 5.4(c) and the watermarked host image has been shown in figure 5.4(d). The value of arbitrary constant a has been chosen such that it is visibly hard to distinguish the difference between the host image and the watermarked image. The watermark is recovered

Figure 5.4. (a) The host image, (b) watermark, (c) encrypted image of watermark with FRT order $\alpha_1 = 0.25$, $\alpha_2 = 0.55$, (d) watermarked host image, and (e) recovered watermark obtained after use of correct RPMs and FRT orders.

successfully after the use of respective correct RPMs and FRT orders as shown in figure 5.4(e). A similar method can be applied for watermarking binary and color data/images.

Further, the watermarking of 3D objects using FRT in DRPE scheme has been reported [23]. In digital holography, the hologram is recorded via a CCD camera and the reconstruction is carried out numerically. Digital storage of the hologram enables one to apply image processing tools for noise removal and reconstructing the object with arbitrary views. Therefore, it is possible that an original digital hologram can be numerically propagated in free space and a new hologram can be created. In this case, the original holographer may lose his right if someone generates a new hologram and claims the ownership. This might allow illegal or unauthorized use of a digital hologram, which can be termed as *Hologram Piracy*. Therefore, it is important to protect the individual's right. Embedding a watermark into the digital hologram could be a potential solution for this possible threat [23]. A watermark may be embedded into the hologram plane but in that case, if a new hologram is created then it will not be possible to recover the watermark without an exhaustive search. For better security purposes, the watermark can be encoded in a plane, which is at a known distance from the object.

A watermarking algorithm employing the FRT and random phase encoding has been reported, which exploits the benefits of FRT [24]. Further, a hierarchical encrypted image watermarking in the FRT domain has been reported, in which four different watermarks were used. Each of the watermarks was encrypted independently in the DRPE scheme and all the encrypted functions were multiplexed. The multiplexed data was embedded into a host image. In this study, multiple watermarks were encrypted with multiple levels of security [25]. To indicate the degree of transparency, the PSNR is an important parameter, which is calculated between the original and retrieved watermarks. To check the robustness of any watermarking scheme, the effects of common image processing operations such as compression, cropping, filtering, noise and occlusion are studied. For authentication of the watermark, CC value is evaluated between the original and the retrieved watermarks. Using non-cascade iterative encrypted kinoform and corresponding the embedding algorithm, a watermarking technique has been reported, in which the kinoform was embedded to two-level discrete wavelet coefficients of the cover image [26]. The scheme claimed to possess high security and resisted malicious attacks.

5.3 Optical asymmetric watermarking

One particular problem with the watermarking techniques is that they are mostly symmetric. It means that the keys used for watermark embedding and retrieval are the same. Therefore, the intended user who is allowed to detect the watermark acquires the knowledge of the parameters used to remove it. Thus, the personnel responsible for embedding watermarks acquire the knowledge of all the critical parameters of the watermarking scheme. A potential solution to this problem could be making the watermarking technique asymmetric in nature. The scheme should use different keys for encoding and retrieval processes. Based on the type of keys used, the watermarking is categorized into symmetric and asymmetric watermarking. In symmetric watermarking, the keys used for watermark embedding and retrieval are the same while in the asymmetric watermarking technique, the keys used for watermark embedding and retrieval are different. Asymmetric watermarking is based on a one-way relationship between the embedding (secret) key and the retrieval (asymmetric) key.

The main issue with the asymmetric watermarking is the detector's design. The detector should not reveal sufficient information, which may lead to the erasure of the embedded watermark, even if the intruder knows the detection algorithm and public detection keys. The design of an asymmetric watermarking scheme is a challenge because revealing too much information would imply the risk of hacking, whereas not revealing too much information might imply less reliable detection. In asymmetric watermarking, the watermark is detected by a correlation test. The reference watermark and the embedded watermark both should have some correlation [27–29].

Employing optical technology, an asymmetric watermarking scheme through an amplitude- and phase-truncation approach has been reported [29]. In most of the encryption algorithms, the encryption keys are in the form of a noise-like distribution, which have a uniform distributed histogram. The noise-like distribution could be an apparent sign for the indication of the presence of the keys and might attract hackers' attention. This may lead to a security problem, while transferring data through some communication channels. To address this potential issue, it is required to transfer the keys to some other meaningful images, which will disguise the attackers. In most of the iterative encryption schemes, support constraints play an important role in decryption. Addressing this issue, in a further study, an asymmetric watermarking technique based on modified wavelet fusion and diffractive imaging has been reported where the support constraints were transferred into different meaningful images captured by the axial translation of the sensor and using the amplitude- and phase-truncation approach. Further, a watermark was embedded into the meaningful images to address the copyright issue [29]. A detailed discussion on the asymmetric cryptosystem is available in chapter 10.

While implementing a practical cryptosystem based on opto-electronic techniques, it is feasible to combine watermarking so that not only information is secured but ownership or copyright is also protected.

References

[1] Kundur D and Hatzinakos D 1998 Improved robust watermarking through attack characterization *Opt. Express* **3** 485–90

[2] Wang H J M, Su P C and Jay Kuo C C 1998 Wavelet-based digital image watermarking *Opt. Express* **3** 491–6

[3] Wang Y, Deherty J F and vanDyck R E 2002 A wavelet-based watermarking algorithm for ownership verification of digital images *IEEE Trans. Image Process.* **11** 77–88

[4] Kumari B P and Rallabandi V P S 2008 Modified patchwork-based watermarking scheme for satellite imagery *Sig. Process.* **88** 891–904

[5] Takai N and Mifune Y 2002 Digital watermarking by a holographic technique *Appl. Opt.* **41** 865–73

[6] Kishk S and Javidi B 2002 Information hiding technique with double phase encoding *Appl. Opt.* **41** 5462–70

[7] Kishk S and Javidi B 2003 Watermarking of three-dimensional objects by digital holography *Opt. Lett.* **28** 167–9

[8] Kishk S and Javidi B 2003 3D watermarking by a 3D hidden object *Opt. Express* **11** 874–88

[9] He M Z, Cai L Z, Liu Q, Yang X C and Wang X F 2005 Multiple image encryption and watermarking by random phase matching *Opt. Commun.* **247** 29–37

[10] Abookasis D, Montal O, Abramson O and Rosen J 2005 Watermarks encrypted in a concealogram and deciphered by a modified joint transform correlator *Appl. Opt.* **44** 3019–23

[11] He M Z, Cai L Z, Liu Q and Yang X L 2005 Phase-only encryption and watermarking based on phase-shifting interferometry *Appl. Opt.* **44** 2600–6

[12] Zhou X, Chen L and Shao J 2005 Investigation of digital hologram watermarking with double binary phase encoding *Opt. Eng.* **44** 067007

[13] Chang H T and Tsan C L 2005 Image watermarking by use of digital holography embedded in the discrete-cosine-transform domain *Appl. Opt.* **44** 6211–9

[14] Liu Z, Xu L, Guo Q, Lin C and Liu S 2010 Image watermarking by using phase retrieval algorithm in gyrator transform domain *Opt. Commun.* **283** 4923–7

[15] Singh N and Sinha A 2010 Digital image watermarking using gyrator transform and chaotic maps *Optik* **121** 1427–37

[16] Javidi B (ed) 2005 *Optical and Digital Techniques for Information Security* (Berlin: Springer)

[17] Liu S, Hennelly B M, Guo C and Sheridan J T 2015 Robustness of double random phase encoding spread-space spread-spectrum watermarking technique *Signal Process.* **109** 345–61

[18] Lu Y, You S, Zhang W, Yang B, Peng R and Zhuang S 2016 Watermarking scheme for microlens-array-based four-dimensional light field imaging *Appl. Opt.* **55** 3397–404

[19] Yadav A K, Vashisth S, Singh H and Singh K 2015 A phase-image watermarking scheme in gyrator domain using devil's vortex Fresnel lens as a phase mask *Opt. Commun.* **344** 172–80

[20] Rajput S K, Kumar D and Nishchal N K 2014 Photon counting imaging and polarized light encoding for secure image verification and hologram watermarking *J. Opt.* **16** 125406

[21] Nishchal N K 2009 Optical image watermarking using fractional Fourier transform *J. Opt. (Springer-India)* **38** 22–8

[22] Kim H and Lee Y H 2005 Optimal watermarking of digital hologram of 3D objects *Opt. Express* **13** 2881–6

[23] Nishchal N K, Pitkaaho T and Naughton T J 2010 Digital Fresnel hologram watermarking *Proc. 9th Euro-American Workshop on Information Optics (Helsinki, Finland, July 11–16, 2010)*

[24] Gu Q, Liu Z and Liucora S 2011 Image watermarking algorithm based on fractional Fourier transform and random phase encoding *Opt. Commun.* **284** 3918–23

[25] Nishchal N K 2011 Hierarchical encrypted image watermarking using fractional Fourier domain random phase encoding *Opt. Eng.* **50** 097003

[26] Deng K, Yang G and Xie H 2011 A blind robust watermarking scheme with non-cascade iterative encrypted kinoform *Opt. Express* **19** 10241–51

[27] Kim T Y, Choi H, Lee K and Kim T 2004 An asymmetric watermarking system with many embedding watermarks corresponding to one detection watermark *IEEE Signal Process. Lett.* **11** 375–7

[28] Fu Y-G 2012 Asymmetric watermarking scheme based on shuffling *Procedia Eng.* **29** 1640–4

[29] Mehra I and Nishchal N K 2014 Optical asymmetric watermarking using modified wavelet fusion and diffractive imaging *Opt. Lasers Eng.* **68** 74–82

IOP Publishing

Optical Cryptosystems

Naveen K Nishchal

Chapter 6

Polarization encoding

6.1 Introduction

Polarization is defined as variation in amplitude and direction of the electric field vector of an optical wave field in one cycle. The different states of polarization are the characteristics of transverse waves, which specify the geometric orientation of oscillations. Spatially encoding or mapping the state of polarization of an incident or imaged coherent light beam has been a subject of interest. Controlling the polarization states of input light is fundamental to several applications, such as information security, data storage, nonlinear optics, imaging, and surface plasmon-based photonic devices. Theories of polarized radiation interaction with optical elements are divided into two groups: Jones calculus and Mueller calculus [1].

Jones calculus
Jones calculus was invented by R Clark Jones in 1941. It is used to describe the effect of an optical device or medium on the polarization state of light. It is based on the basic assumption of coherent addition of waves, whose theoretical analysis begins with Maxwell's equations. Since light is an electromagnetic wave, this method uses the most natural way of representing the light in terms of the electric field vector. In general, this method deals with an instantaneous field, and thus preferred to describe an optical system with coherent sources such as lasers. However, the Jones approach is not suitable for extracting information associated with depolarizing effects.

The polarization state of light, described by a two-element Jones vector, comprises instantaneous components of an electric field as represented by $E_x(t)$ and $E_y(t)$. In general, these components are complex and carry both amplitude and phase information.

$$E_{\text{in}} = \begin{bmatrix} E_x(t) \\ E_y(t) \end{bmatrix} \tag{6.1}$$

doi:10.1088/978-0-7503-2220-1ch6

The effect on polarization state of light after interaction with an optical device (OD) can be described by the following linear operation:

$$E_{\text{out}} = JE_{\text{in}} \tag{6.2}$$

The coupling between vector E_{in} and E_{out}, representing incident and output light field respectively, can be described by a set of four coefficients representing the Jones matrix, J of the optical device. The properties of several optical devices acting in series can be described through a single matrix simply by the Jones matrices of the devices.

Mueller calculus

The polarization state of light and its interaction with the optical medium can be alternately described by Mueller calculus. This technique was developed in 1943 by Hans Mueller. The method can describe a more general situation of a linear optical system as compared to the description applicable for the Jones calculus. The basic assumption of Mueller calculus is the incoherent addition of light waves. For theoretical analysis, the relation between input and output Stokes vector is considered to be linear. The development of this method is heuristic and lacks the mathematical rigor of Jones calculus. This method directly deals on the Stokes vector rather than field vectors. The associated Stokes parameters are based on intensity measurements and are directly measureable. The Stokes parameters describe a time-averaged optical signal and therefore, the Mueller method is often preferred in a situation where the polarization state of light changes rapidly and randomly, such as natural sunlight.

Disregarding the coherent wave superposition, any fully polarized, partially polarized, or un-polarized light can be represented by four Stokes parameters S_0, S_1, S_2, and S_3. The Stokes column vector is represented as

$$S_{\text{in}} = \begin{bmatrix} S_0 \\ S_1 \\ S_2 \\ S_3 \end{bmatrix} \tag{6.3}$$

A real-valued matrix M, characterized by the properties of the optical element acts on the input polarization state S_{in} by matrix multiplication to produce output state S_{out}.

$$S_{\text{out}} = M \, S_{\text{in}} \tag{6.4}$$

As indicated by equation (6.4), the output Stokes parameters are a linear combination of the input Stokes parameters, each parameter weighted by the corresponding elements of Mueller matrix, M. An optical medium can be mathematically represented by a 4×4 Mueller matrix, which acts on the Stokes vector. It has 16 elements and it does not contain phase information. The Mueller matrix, M, is defined as

$$M = \begin{bmatrix} m_{00} & m_{01} & m_{02} & m_{03} \\ m_{10} & m_{11} & m_{12} & m_{13} \\ m_{20} & m_{21} & m_{22} & m_{23} \\ m_{30} & m_{31} & m_{32} & m_{33} \end{bmatrix} \tag{6.5}$$

Optical techniques for image encryption have gained much attention because of inherent advantages such as parallel processing, high space-bandwidth product, and multi-dimensional capability. Polarization of light has been used to encode and store image/data. States of polarization are considered as one of the significant properties of the light waves that provide additional flexibility in the encryption key design and also in the encryption process. Manipulation of states of polarization with amplitude and phase modulation helps to achieve better key design. Polarization-based security systems have an added advantage over the DRPE schemes that only intensity is recorded while the output of DRPE is always complex and recording poses a practical problem. Due to intensity, recording such methods offers ease in storage and transmission without compromising the security. Due to the commercial availability of amplitude and phase-only SLMs, a large number of demonstrations for information encoding based on polarization have been reported in the literature.

Broadly speaking, the polarization-based encryption can be categorized into two approaches: the spatially modulated retardation approach and spatially modulated azimuthal angle approach. The mathematical analysis is carried out by either Jones calculus or Mueller calculus. In Jones (complex) calculus based analysis, three different intensity measurements are required while in Mueller (real) calculus, four different intensity measurements are required [1]. Mueller formalism has an advantage over the Jones analysis whereby birefringent properties of the elements are not required for the analysis process. Therefore, Mueller formalism is insensitive to spatial fabrication errors and can be implemented with incoherent, polychromatic, and unpolarized light beams.

6.2 Double random phase polarization encoding

In most of the reported studies, encoding has been achieved with linear polarization states having different azimuths. Also, mostly binary data has been considered as inputs because they can be easily encoded as orthogonal polarizations. This section reviews some of the reported literature on polarization encoding-based security systems.

In the year 2000, a phase-only encryption-decryption using polarization sensitive element was reported. The encryption was achieved by changing the relative phases of light in a 2D wavefront and polarization was used for the visualization of the decoded information. The pixel mapping approach was implemented through two parallel aligned liquid crystal SLMs [2]. In this polarization encoding method, each pixel of the SLM acted as a voltage-controlled wave plate which can be programmed over a 2π phase range. For changing the state of polarization of a laser beam, two systems were reported. The first method involved rotation of the major axis of an elliptically polarized beam by an arbitrary angle and in the second method, a linearly polarized input beam was converted into an arbitrary elliptically polarized beam. Further, by combining two systems an arbitrary state of polarization was produced [3].

A polarization-based security verification system has been reported, in which polarization-encoded data was used in a nonlinear JTC architecture [4]. The polarization-encoded image looks similar to the original image because states of polarization cannot be detected by an intensity sensitive detector. A polarization-encoded image was compared with a reference polarization mask. An encryption system for securing a binary image was also demonstrated, in which an exclusive-OR (XOR) operation was performed between the image and an RPM. The XOR operation was implemented by using two ferroelectric liquid crystal SLMs. The encoded information as polarization states was successfully converted to intensity variations through a polarizer [5]. The study was further extended with cascaded SLMs [6]. A system for controlling the relative phase of each modulation cell in the SLM has been demonstrated with two optically addressed phase-only SLMs. In this method, the state of polarization was spatially encoded into an elliptically polarized wavefront, controlling both ellipticity and the rotation angle [7]. For modulating the polarization state of an optical wave, it is necessary that the wave should propagate through a birefringent medium so that the relative phase between two orthogonal wave components can be changed.

A secured holographic memory based on polarization encoding has been reported, in which original data was represented as polarization states. Using a polarization modulation mask in the input plane, the original polarization distribution was encoded into a random polarization state and was recorded in a polarization sensitive material. The mask can rotate the direction of the principal axis of the elliptically polarized light and can also modulate the phase retardation at each pixel [8]. Since geometrical phases originate from the polarization state manipulation, this approach has been used for polarization encryption [9, 10]. Also, computer-generated space-variant subwavelength gratings (SWG) have been used in manipulation of the complex polarization states. The SWG imprints the image intensity along with the random key function in the local orientation of the waveplate's fast axis. For decryption, the encrypted element has to be illuminated with circularly polarized light. Retrieval of the original image is possible after analysis of the Stokes parameters.

An optical system capable of encoding 2D signals as distinct polarization states has been reported. In this system, for transforming a single input signal to a polarization space, a single SLM has been used. For transforming two input signals, two SLMs can be used. Considering each input signal assumes N values and the system generates distinct polarization states for each N^2 combination, then the resulting output signal is considered as a multiplexed version [11]. The optical system consisted of SLM, linear polarizers, and waveplate retarders, in which the SLM has been used as a programmable birefringent device. The possible application of the scheme could be in secure holographic storage combined with polarization multiplexing. A method for encoding binary data into circular polarization states with opposite helicity has been demonstrated [12]. The method produced a steganographic imaging system in which binary data was hidden on a background with the same state of polarization but opposite helicity. A detailed study on the Jones matrix method for the analysis of coherent optical Fourier processor employing structured polarization has been reported [13]. A Jones matrix was derived that can describe polarization output in terms of two vectorial functions.

Polarization of light using Stokes–Mueller calculus has an advantage over Jones vector calculus whereby it manipulates only intensity information. Exploring this property, a dual encryption scheme using the polarization of light has been reported [14]. In this scheme, Stokes–Mueller formalism was used to parameterize the intensity images. The scheme was analyzed to chosen ciphertext attack, chosen-plaintext attack, known-plaintext attack, brute force attack, and video sequence attack [15]. The scheme showed resistance against brute force and video sequence attacks because of the high number of combinations for the keys but was vulnerable to known- and chosen-plaintext attacks when more than one plaintext-ciphertext pair was known.

Furthering the use of polarization in information security, the generation of arbitrary 3D polarization orientation with a vectorial beam has been reported [16]. Implementing this method to gold nanorods, orientation-unlimited polarization encryption was demonstrated. The method enabled polarization encryption with a 3D arbitrary key. The 3D polarization orientation is essential to the interaction between 3D physically anisotropic micro- and nano-objects. In free space, the wave nature of light confines the polarization orientation control only to the planar regimes, therefore a method was required to break this limit. The wavefront of light was bent through vectorial diffraction by a high numerical aperture lens. The vectorial diffraction of the superposed beam (superposition of radially and azimuthally polarized beams) constructed the arbitrary 3D polarization orientation. This method showed lots of potential and could be used in various applications. Further, 3D polarization multiplexing by optimizing a vectorial beam using a multiple-signal window plane phase retrieval algorithm has been reported [17]. In this proposal, input messages were represented with multiple QR codes, which were partitioned into a series of sub-blocks and each sub-block was marked with a specific polarization state. Optical image encryption based on interference between two polarized wavefronts has also been reported [18]. In this technique, a polarization selective diffractive optical element (PSDOE), which is a phase-only optical element, was used to generate the desired polarized wavefronts. The PSDOE can be fabricated on a birefringent substrate having ordinary and extra-ordinary polarized light with different refractive indices.

Since polarization-based schemes have been found to be vulnerable to various types of attacks [15], a polarization-based asymmetric cryptosystem has been reported [19, 20]. An asymmetric color cryptosystem using PSDOE and SPM has also been reported, in which the concept of an amplitude- and phase-truncation approach was combined with the interference of polarized wavefronts. The method has been applied to single-channel color image encryption using FRT domain SPM encoding, which alleviated the alignment problem of interference but maintained a high level of security without using iterative encoding [19]

Using Stokes–Mueller formalism, polarized light encoding has been reported, in which an input image bonded with an RPM was Fresnel transformed, which resulted in a complex spectrum. The obtained spectrum was amplitude-truncated and phase-truncated. The phase-truncated part was encrypted using polarized light by using two independent optical plane waves. The first plane wave illuminated the input image and encoded it into a defined polarization state and the second plane wave illuminated the intensity key image and encoded into another polarization state. These two waves were multiplexed, which resulted in the first level of encryption. The resulting waves were then passed through a matrix of a linear polarizer to obtain the second level of encryption. For decryption, the whole process was reversed applying the correct keys. The study was used for gray-scale and color image encryption-decryption. Due to the generation of asymmetric keys, the method showed immunity against special attack. The study has been further combined with photon counting imaging for encryption and authentication of digital hologram watermarking [21]. Also, an asymmetric system based on elliptical polarized light linear truncation and numerical reconstruction has been reported [22]. To achieve linear truncation on the spatially resolved elliptical polarization distribution, an array of linear polarizers were used.

A polarimetric image encryption and verification has been reported, which used random polarized vector keys. The polarization information of the encrypted signal was computed from the Stokes parameters. The concept of photon-counting was combined for generating sparse data for verification purposes [23]. The selection of the polarization of the input light beam is a key element in an encryption system based on highly focused fields. The analysis of the performance of such a system's dependence on the polarization state of the input beam has also been studied. It has been stated that an encryption system using quantum techniques shows stronger strength against attacks [24].

A security scheme using a vector beam characterized by spatially variant polarization distribution has been reported. A vector beam having helical components carrying tailored phases was generated corresponding to an image to be encrypted. The modified Gerchberg–Saxton phase retrieval algorithm was used for tailoring the phase. Stokes parameters for the final vector beam was used for constructing the ciphertext and one of the security keys. The proposed scheme generated real ciphertext and keys, offering a benefit in easy transmission and storage [25]. An encryption system based on polarization encoding employing incoherent imaging has also been reported [26]. In this method, incoherent point spread function of the imaging system served as the key for encrypting data. The analysis was based on Mueller calculus.

6.3 Polarization encoding-based asymmetric cryptosystem

In this section, an image encryption process using amplitude and phase truncation in Fresnel domain with polarized light is discussed [20]. An input image to be encrypted, $f(x,y)$, bonded with an RPM is Fresnel transformed. Consider that λ

and d denote wavelength and free space propagation distance, respectively. The obtained Fresnel spectrum is expressed as

$$E(u,v) = \frac{\exp\left\{\frac{i2\pi d}{\lambda}\right\}}{\sqrt{i\lambda d}} \iint \{f(x,y) \times \exp[i2\pi R(x,y)]\}$$
$$\times \exp\left[\frac{i\pi}{\lambda d}((x-u)^2 + (y-v)^2)\right]dxdy \tag{6.6}$$

The obtained complex function is amplitude-truncated (AT) and phase-truncated (PT) as

$$E_P(u,v) = \text{AT}\{E(u,v)\} \tag{6.7}$$

$$E_A(u,v) = \text{PT}\{E(u,v)\} \tag{6.8}$$

Using Stokes–Mueller calculus, the phase-truncated data, $E_A(u,v)$, is parameterized. Consider the Mueller matrix of a linear polarizer with the orientation of its transmission axis, φ_P, is $M_{\text{Pol}}(\varphi_P)$.

$$M_{\text{Pol}}(\varphi_P) = \frac{1}{2}\begin{bmatrix} 1 & \cos(2\varphi_P) & \sin(2\varphi_P) & 0 \\ \cos(2\varphi_P) & \cos^2(2\varphi_P) & \frac{1}{2}\sin(4\varphi_P) & 0 \\ \sin(2\varphi_P) & \frac{1}{2}\sin(4\varphi_P) & \sin^2(2\varphi_P) & 0 \\ 0 & 0 & 0 & 0 \end{bmatrix} \tag{6.9}$$

The Mueller matrix of a linear retarder, with phase shift φ_R and fast-axis orientation θ, is $M_{\text{Re}\,t}(\varphi_R,\theta)$.

$$M_{\text{Re}\,t}(\varphi_R,\theta) = P(-\theta)\begin{bmatrix} 1 & 0 & 0 & 0 \\ 0 & 1 & 0 & 0 \\ 0 & 0 & \cos(\varphi_R) & \sin(\varphi_R) \\ 0 & 0 & -\sin(\varphi_R) & \cos(\varphi_R) \end{bmatrix}\text{Rot}(\theta) \tag{6.10}$$

where $\text{Rot}(\theta)$ is the 4×4 rotation matrix,

$$\text{Rot}(\theta) = \begin{bmatrix} 1 & 0 & 0 & 0 \\ 0 & 1 & \cos(\theta) & \sin(\theta) \\ 0 & 0 & -\sin(\theta) & \cos(\theta) \\ 0 & 0 & 0 & 0 \end{bmatrix} \tag{6.11}$$

Consider the Stokes vector for the phase-truncated function, $E_P(u,v)$, and the intensity key image are denoted as S_P and S_k, respectively. For any unpolarized light, Stokes vector can be written as

$$S^{(i,j)} = \begin{bmatrix} S_0^{(i,j)} \\ 0 \\ 0 \\ 0 \end{bmatrix} \tag{6.12}$$

Assuming Stokes vectors of a phase-truncated image and key image are unpolarized initially, then the Stokes vectors can be expressed as

$$S_P^{(i,j)} = \begin{bmatrix} S_{P0}^{(i,j)} & 0 & 0 & 0 \end{bmatrix}^T \tag{6.13}$$

$$S_k^{(i,j)} = \begin{bmatrix} S_{k0}^{(i,j)} & 0 & 0 & 0 \end{bmatrix}^T \tag{6.14}$$

The subscripts P and k stand for phase-truncated and key images, respectively. Here, T denotes the transpose of a matrix. The polarized Stokes vectors for the phase-truncated image and key image can be written as

$$S_P^{'(i,j)} = M_{\mathrm{Re}\,t}(\varphi_R,\theta)M_{\mathrm{Pol}}(\varphi_P)S_P^{(i,j)} = \begin{bmatrix} \alpha_P S_{P0}^{(i,j)} & \beta_P S_{P0}^{(i,j)} & \gamma_P S_{P0}^{(i,j)} & \delta_P S_{P0}^{(i,j)} \end{bmatrix}^T \tag{6.15}$$

$$S_k^{'(i,j)} = M_{\mathrm{Re}\,t}(\varphi_R,\theta)M_{\mathrm{Pol}}(\varphi_P)S_k^{(i,j)} = \begin{bmatrix} \alpha_k S_{k0}^{(i,j)} & \beta_k S_{k0}^{(i,j)} & \gamma_k S_{k0}^{(i,j)} & \delta_k S_{k0}^{(i,j)} \end{bmatrix}^T \tag{6.16}$$

The coefficients α, β, γ, and δ depend on φ_R, θ, and φ_P. Their relationship can be explained through the following expressions:

$$\alpha = 1/2 \tag{6.17}$$

$$\beta = 1/2\{\cos(2\theta)\cos[2(\theta - \varphi_P)] + \cos(\varphi_R)\sin[2(\theta - \varphi_P)]\} \tag{6.18}$$

$$\gamma = 1/2\{\sin(2\theta)\cos[2(\theta - \varphi_P)] - \cos(\varphi_R)\sin[2(\theta - \varphi_P)]\} \tag{6.19}$$

$$\delta = 1/2\,\sin(\varphi)\sin[2(\theta - \varphi_P)] \tag{6.20}$$

The polarized Stokes vectors for phase-truncated and key images are multiplexed. In the actual experiment, a beam splitter would combine the light waves passed through the phase-truncated image and key image. A schematic diagram is shown in figure 6.1. The multiplexed output resultant Stokes vector can be expressed as

$$S_R^{'(i,j)} = \begin{bmatrix} \alpha_P S_{P0}^{(i,j)} + \alpha_k S_{k0}^{(i,j)} & \beta_P S_{P0}^{(i,j)} + \beta_k S_{k0}^{(i,j)} & \gamma_P S_{P0}^{(i,j)} + \gamma_k S_{k0}^{(i,j)} & \delta_P S_{P0}^{(i,j)} + \delta_k S_{k0}^{(i,j)} \end{bmatrix}^T \tag{6.21}$$

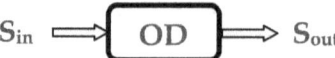

Figure 6.1. Effect of an optical device (OD) on the polarization of light. Stokes vectors input and output states are denoted by S_{in} and S_{out}.

After passing through pixilated polarizer $M_{\text{Pol}}\left(\psi_{\text{random}}^{(i,j)}\right)$, the Stokes vector $S_R^{'(i,j)}$ becomes

$$S_c^{'(i,j)} = \begin{bmatrix} P_{c0}^{(i,j)} & P_{c1}^{(i,j)} & P_{c2}^{(i,j)} & 0 \end{bmatrix}^T \tag{6.22}$$

From the Stokes vector as given by equation (6.22), three encrypted images can be obtained. They are expressed as

$$P_{E_0}^{(i,j)} = \frac{1}{2}\Big[\left(\alpha_P S_{P0}^{(i,j)} + \alpha_k S_{k0}^{(i,j)}\right) + \left(\beta_P S_{P0}^{(i,j)} + \beta_k S_{k0}^{(i,j)}\right) \times \cos\left(2\varphi_{P\text{random}}^{(i,j)}\right)$$
$$+ \left(\gamma_P S_{P0}^{(i,j)} + \gamma_k S_{k0}^{(i,j)}\right) \times \sin\left(2\varphi_{P\text{random}}^{(i,j)}\right)\Big] \tag{6.23}$$

$$P_{E_1}^{(i,j)} = \frac{1}{2}\Big[\left(\alpha_P S_{P0}^{(i,j)} + \alpha_k S_{k0}^{(i,j)}\right) \times \cos\left(2\varphi_{P\text{random}}^{(i,j)}\right)$$
$$+ \left(\beta_P S_{P0}^{(i,j)} + \beta_k S_{k0}^{(i,j)}\right) \times \cos^2\left(2\varphi_{P\text{random}}^{(i,j)}\right)$$
$$+ \frac{1}{2}\left(\gamma_P S_{P0}^{(i,j)} + \gamma_k S_{k0}^{(i,j)}\right) \times \sin\left(4\varphi_{P\text{random}}^{(i,j)}\right)\Big] \tag{6.24}$$

$$P_{E_2}^{(i,j)} = \frac{1}{2}\Big[\left(\alpha_P S_{P0}^{(i,j)} + \alpha_k S_{k0}^{(i,j)}\right) \times \sin\left(2\varphi_{P\text{random}}^{(i,j)}\right)$$
$$+ \frac{1}{2}\left(\beta_P S_{P0}^{(i,j)} + \beta_k S_{k0}^{(i,j)}\right) \times \sin\left(4\varphi_{P\text{random}}^{(i,j)}\right)$$
$$+ \left(\gamma_P S_{P0}^{(i,j)} + \gamma_k S_{k0}^{(i,j)}\right) \times \sin^2\left(2\varphi_{P\text{random}}^{(i,j)}\right)\Big] \tag{6.25}$$

The term $\varphi_{P\text{random}}^{(i,j)}$ represents the angle of the pixilated polarizer. Its value can be chosen randomly between $-\pi$ and $+\pi$. For decryption the process is reversed. The encrypted image has to pass through another pixilated polarizer with appropriate angles (figure 6.2). For the phase-truncated part, the Stokes vectors can be obtained through the following equations:

$$S_{P_0}^{(i,j)} = \frac{2P_{E_0}^{(i,j)} - S_{k0}^{(i,j)}\left\{\alpha_k + \beta_k \cos\left(2\varphi_{P\text{random}}^{(i,j)}\right) + \gamma_k \sin\left(2\varphi_{P\text{random}}^{(i,j)}\right)\right\}}{\alpha_P + \beta_P \cos\left(2\varphi_{P\text{random}}^{(i,j)}\right) + \gamma_P \sin\left(2\varphi_{P\text{random}}^{(i,j)}\right)} \tag{6.26}$$

$S_{P_0}^{(i,j)}$

$$= \frac{2P_{E_1}^{(i,j)} - S_{k0}^{(i,j)}\left\{\alpha_k \cos\left(2\varphi_{P\text{random}}^{(i,j)}\right) + \beta_k \cos^2\left(2\varphi_{P\text{random}}^{(i,j)}\right) + \frac{1}{2}\gamma_k \sin\left(4\varphi_{P\text{random}}^{(i,j)}\right)\right\}}{\alpha_P \cos\left(2\varphi_{P\text{random}}^{(i,j)}\right) + \beta_P \cos^2\left(2\varphi_{P\text{random}}^{(i,j)}\right) + \frac{1}{2}\gamma_P \sin\left(4\varphi_{P\text{random}}^{(i,j)}\right)} \tag{6.27}$$

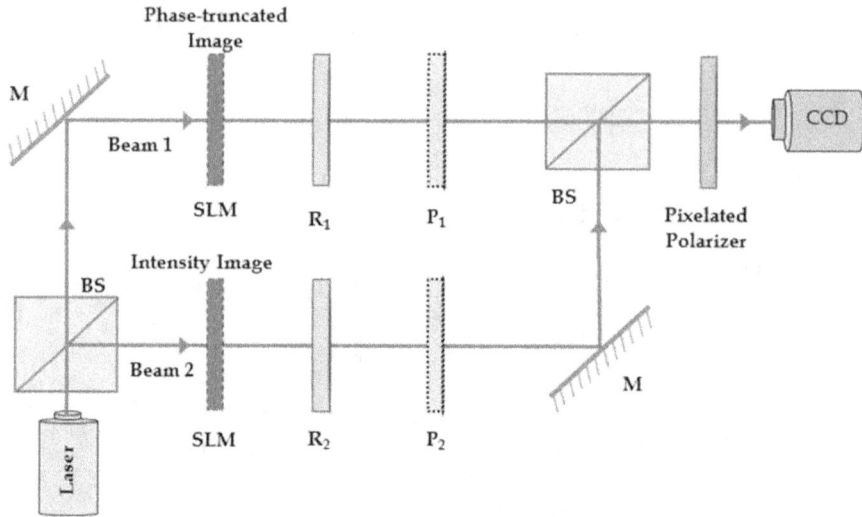

Figure 6.2. Schematic diagram of polarized light-based image encryption scheme. R: retarder; P: polarizer; BS: beam splitter; SLM: spatial light modulator; M: mirror; CCD: charge-coupled device camera.

$$S_{P_0}^{(i,j)}$$

$$= \frac{2I_{E_2}^{(i,j)} - S_{k0}^{(i,j)}\left\{\alpha_k \sin\left(2\varphi_{P\,\text{random}}^{(i,j)}\right) + \frac{1}{2}\beta_k \sin\left(4\varphi_{P\,\text{random}}^{(i,j)}\right) + \gamma_k \sin^2\left(2\varphi_{P\,\text{random}}^{(i,j)}\right)\right\}}{\alpha_P \sin\left(2\varphi_{P\,\text{random}}^{(i,j)}\right) + \frac{1}{2}\beta_P \sin\left(4\varphi_{P\,\text{random}}^{(i,j)}\right) + \gamma_P \sin^2\left(2\varphi_{P\,\text{random}}^{(i,j)}\right)} \quad (6.28)$$

Each of these Stokes vectors obtained through equations (6.26)–(6.28) is corresponding to the phase-truncated value, $E_A(u,v)$, which is further modulated with the amplitude-truncated value and inverse Fresnel transformed:

$$f(x,y) = PT\{\mathfrak{I}_{\lambda}^{-d}[E_A(u,v) \times E_P(u,v)]\} \quad (6.29)$$

As stated in the introduction section, whereby in the reported literature on polarization encoding, mostly binary data has been considered as inputs because binary data can be encoded as orthogonal polarizations. But an asymmetric cryptosystem based on polarization has been applied for securing gray-scale images [20]. This method has been reported for color image encryption also. Color is considered an effective descriptor. Color image encryption is very important because it provides additional information than gray scale and binary images. The primary color components contribute to the higher level of security.

6.3.1 Color image encryption

As reported in literature, most of the color image encryption schemes belong to the three-channel cryptosystems. In such a scheme, a color image is separated into their primary components (red, green, and blue), and correspondingly each color component is encrypted independently which is defined as one channel. Thus, all

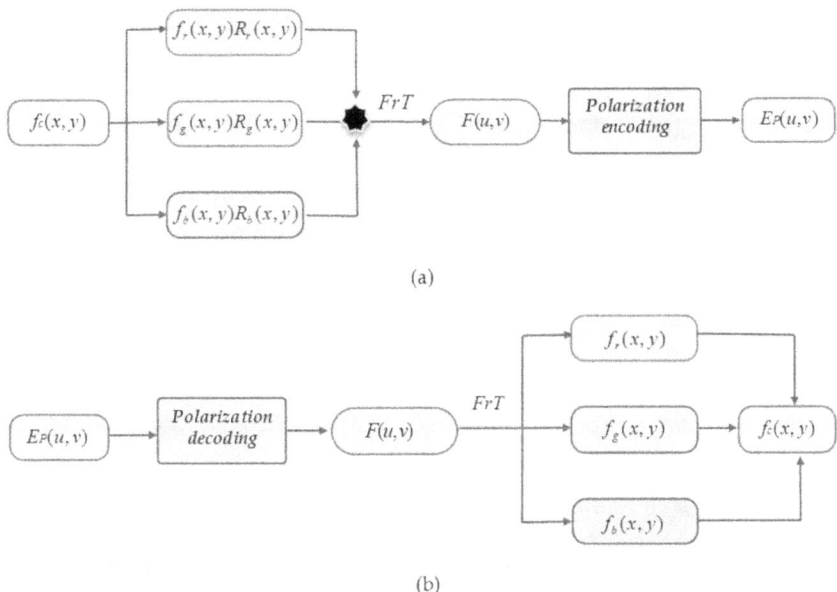

(a)

(b)

Figure 6.3. Block diagram of single-channel color image for (a) encryption and (b) decryption.

the three channels are multiplexed together to get a single encrypted image. For decryption, the correct keys are appropriately applied in the reverse process to retrieve the individual color components, which upon recombination results in the original color image. This technique has been reported under the DRPE framework and also under asymmetric schemes [27]. Since a three-channel encryption system has increased complexity and extra processing cost, single-channel color image encryption techniques have been proposed [28–32]. Combining chaos and FRT, a single-channel color image encryption scheme has also been proposed [30].

Figure 6.3 shows the block diagram for single-channel color image encryption-decryption procedure using Stokes–Mueller formalism. The color image to be encrypted, $f_c(x,y)$, is separated into r (red), g (green), and b (blue) components. Each component is multiplied with an RPM.

$$\left.\begin{aligned}
f_{rm}(x,y) &= f_r(x,y)R_r(x,y) \\
f_{gm}(x,y) &= f_g(x,y)R_g(x,y) \\
f_{bm}(x,y) &= f_b(x,y)R_b(x,y)
\end{aligned}\right\} \tag{6.30}$$

Thus all the three modulated obtained components are expressed as [32]

$$f_m(x,y) = f_{rm}(x,y) \otimes f_{gm}(x,y) \otimes f_{bm}(x,y) \tag{6.31}$$

where $\otimes$ denotes the convolution operation. The function, $f_m(x,y)$ is Fresnel transformed as

$$F(u,\upsilon) = \mathfrak{I}_\lambda^d[f_m(x,y)] \tag{6.32}$$

Applying convolution theorem, equation (6.29) can be expressed as

$$F(u,v) = F_{rm}(u,v) \times F_{gm}(u,v) \times F_{bm}(u,v) \tag{6.33}$$

where $F_{rm}(u,v)$, $F_{gm}(u,v)$, and $F_{bm}(u,v)$ denote the Fresnel transforms of $f_{rm}(x,y)$, $f_{gm}(x,y)$, and $f_{bm}(x,y)$, respectively. For generating asymmetric keys, the obtained spectrum, $F(u,v)$, is amplitude- and phase-truncated. The truncation process as expressed in equations (6.8) and (6.9) is repeated.

$$F_P(u,v) = AT[F(u,v)] \tag{6.34}$$

$$F_A(u,v) = PT[F(u,v)] \tag{6.35}$$

The phase-truncated value, $F_A(u,v)$, is encrypted using Stokes–Mueller formalism. The process explained through equations (6.10)–(6.25) is repeated. The phase-truncated value $F_A(u,v)$ can be rewritten as

$$F_A(u,v) = |F_{rm}(u,v)| \cdot |F_{gm}(u,v)| \cdot |F_{bm}(u,v)| \tag{6.36}$$

Here, $|\cdot|$ denotes the modulus operation. The amplitude-truncated values corresponding to each color component help generate asymmetric keys.

$$D_r(u,v) = \frac{AT[F_{rm}(u,v)]}{|F_{gm}(u,v)| \times |F_{bm}(u,v)|} \tag{6.37}$$

$$D_g(u,v) = \frac{AT[F_{gm}(u,v)]}{|F_{rm}(u,v)| \times |F_{bm}(u,v)|} \tag{6.38}$$

$$D_b(u,v) = \frac{AT[F_{bm}(u,v)]}{|F_{rm}(u,v)| \times |F_{gm}(u,v)|} \tag{6.39}$$

For decryption, the polarization decoding is carried out. The mathematical steps as expressed in equations (6.26)–(6.28) are followed, which gives the convolved phase-truncated value. This function is multiplied with corresponding keys and its inverse Fresnel transformation is obtained, which retrieves the original color component.

$$f_r(x,y) = \mathfrak{F}_\lambda^{-d}[F_A(u,v) \times D_r(u,v)] \tag{6.40}$$

$$f_g(x,y) = \mathfrak{F}_\lambda^{-d}[F_A(u,v) \times D_g(u,v)] \tag{6.41}$$

$$f_b(x,y) = \mathfrak{F}_\lambda^{-d}[F_A(u,v) \times D_b(u,v)] \tag{6.42}$$

Using equations (6.40)–(6.42), all the three components are decrypted successfully, which are then combined to retrieve the original color image.

The decryption can be implemented optically employing the digital holography principle. A schematic diagram has been shown in figure 6.4 [20]. The phase-truncated part is a real function, which can be displayed over an SLM and its Fresnel transformation can be obtained, which will interfere with a plane reference

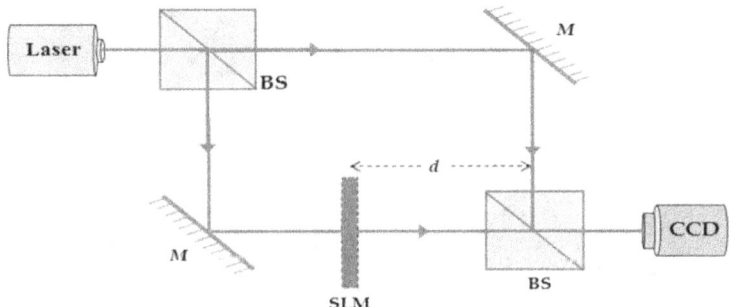

Figure 6.4. Schematic diagram for a phase-truncated Fresnel transform-based decryption scheme.

beam and the interference pattern will be recorded by a CCD camera. Any interferometer can be used for the process, such as a Michelson interferometer or Mach–Zehnder interferometer. The schematic of a Mach–Zehnder interferometer has been shown. The recorded digital hologram can be stored in a personal computer, which can be numerically reconstructed. The key generation part that is the amplitude- and phase-truncation process should be carried out digitally.

The computer simulation results using a MATLAB 7.10 platform for gray-scale and color images have been shown in figures 6.5 and 6.6, respectively. The results have been reproduced with permission from [20]. A gray-scale image of Lena of size 512×512 pixels, as shown in figure 6.5(a), was used as the input image to be encrypted. For Fresnel transform computation, $\lambda = 632$ nm and $d = 165$ mm have been used. The phase-truncated image is shown in figure 6.5(b). Because phase-truncation leaves the real part only, a random intensity mask was used for encryption instead of an RPM, as the key.

This intensity mask is shown in figure 6.5(c). The value of phase shifts for retarders R_1 and R_2 has been used as $\varphi_R = \pi/2$. Thus the obtained encrypted image is shown in figure 6.5(d). The decrypted image obtained after computing Fresnel transform for $\lambda = 632$ nm and $d = -165$ mm along with correct intensity mask is as figure 6.5(a). The decrypted images cannot be obtained after using the wrong intensity mask, wrong amplitude-truncated value, and with a different angle of the pixelated polarizer.

The simulation results for a single-channel color image encryption have been shown in figure 6.6. A color image of Barbara of size $512 \times 512 \times 3$ pixels used for encryption is shown in figure 6.6(a). The color image was separated into primary components: red, green, and blue. These color components were convolved in the Fresnel transform domain and resulted in a single gray-scale image. The convolved image was encrypted using a random intensity mask. When all the correct keys are used, the original image is retrieved. If the keys are not used correctly or wrong keys are used, a noisy pattern appears. The decrypted color images obtained after using the wrong intensity key image, encryption key, and the wrong angle of the pixelated polarizer have been shown in figures 6.6(b)–(d), respectively.

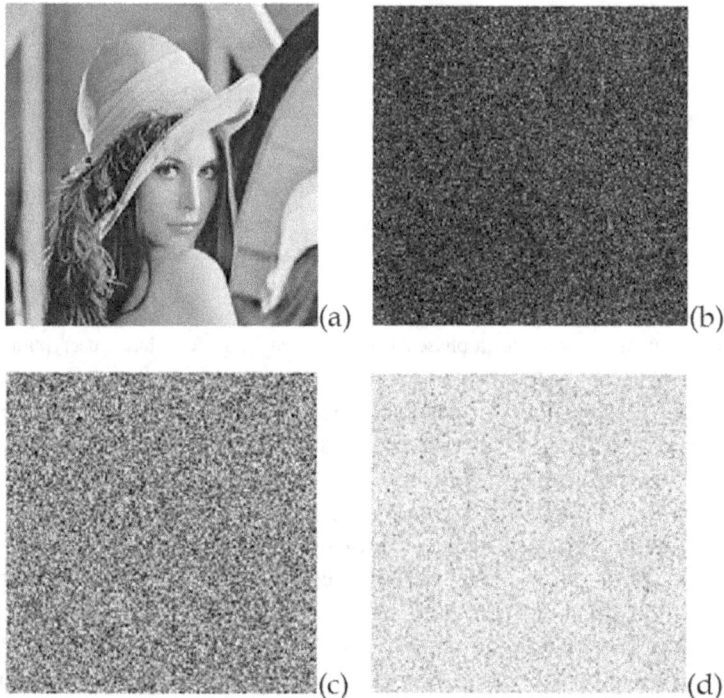

Figure 6.5. Simulation results for a gray-scale image. (a) Lena image, (b) phase-truncated image, (c) random intensity image, and (d) encrypted image. Reprinted with permission from [20]. © The Optical Society.

Usually MSE and correlation coefficient (CC) are calculated between the decrypted image and original image to check the quality of the decryption. These are suitable parameters for binary and gray-scale images. In the case of the color image, the performance of the encryption-decryption system is checked in terms of relative error (RE) and CC values between the original and the decrypted color image.

The RE for a color image is defined as

$$RE_{rgb} = \frac{RE_r + RE_g + RE_b}{3} \qquad (6.43)$$

The computed value of RE should be very low with correct keys and the value should be extremely high when wrong keys are used.

The CC for a color image is defined as

$$CC_{rgb} = \frac{CC_r + CC_g + CC_b}{3} \qquad (6.44)$$

The computed value of CC should be very high with correct keys and the value should be extremely low when wrong keys are used.

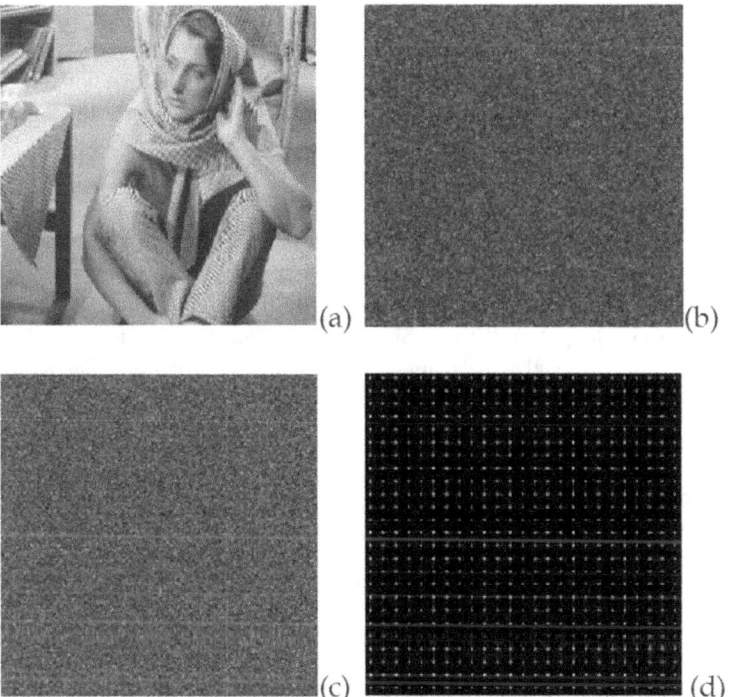

Figure 6.6. Simulation results for a color image. (a) The decrypted color image obtained after using all correct keys, (b) the decrypted color image obtained after using the wrong intensity key image, (c) the decrypted color image obtained after using the encryption key, and (d) the decrypted color image obtained after using the wrong angle of the pixelated polarizer. Reprinted with permission from [20]. © The Optical Society.

Having discussed the cryptosystems employing the polarization of light, it can be inferred that polarization-based security systems can be considered as a potential candidate for practical security devices.

References

[1] Brosseau C 1998 *Fundamentals of Polarized Light: A Statistical Optics Approach* (New York: Wiley)

[2] Mogensen P C and Gluckstad J 2000 A phase image optical encryption system with polarization encoding *Opt. Commun.* **173** 177–83

[3] Davis J A, Mc Namara D E, Cottrell D M and Sonehara T 2000 Two-dimensional polarization encoding with a phase-only liquid crystal spatial light modulator *Appl. Opt.* **39** 1549–54

[4] Javidi B and Nomura T 2000 Polarization encoding for optical security systems *Opt. Eng.* **39** 2439–43

[5] Unnikrishnan G, Pohit M and Singh K 2000 A polarization encoded optical system using ferroelectric spatial light modulator *Opt. Commun.* **185** 25–31

[6] Tu H Y, Cheng C J and Chen M L 2004 Optical image encryption based on polarization encoding by liquid crystal spatial light modulator *J. Opt. A: Pure Appl. Opt.* **6** 524–28

[7] Eriksen R, Mogensen P and Gluckstad J 2001 Elliptical polarization encoding in two dimensions using phase-only spatial light modulators *Opt. Commun.* **187** 325–36

[8] Tan X, Matoba O, Okada-Shudo Y, Ide M, Shimura T and Kuroda K 2001 Secure optical memory system with polarization encryption *Appl. Opt.* **40** 2310–5

[9] Biener G, Niv A, Kleiner V and Hasman E 2005 Geometrical phase image encryption obtained with space-variant subwavelength gratings *Opt. Lett.* **30** 1096–8

[10] Biener G, Niv A, Kleiner V and Hasman E 2006 Space-variant polarization scrambling for image encryption obtained with subwavelength gratings *Opt. Commun.* **261** 5–12

[11] Unnikrishnan G, Naughton T J and Sheridan J T 2006 Polarization encoding and multiplexing of two-dimensional signals: application to image encryption *Appl. Opt.* **45** 5693–700

[12] Martinez J L, Moreno I and Mateos F 2008 Hiding binary optical data with orthogonal circular polarizations *Opt. Eng.* **47** 030504

[13] Moreno I, Iemmi C, Campos J and Yzuel M J 2011 Jones matrix treatment for optical Fourier processors with structured polarization *Opt. Express* **19** 4583–94

[14] Alfalou A and Brosseau C 2010 Dual encryption scheme of images using polarized light *Opt. Lett.* **35** 2185–7

[15] Dubreuil M, Alfalou A and Brosseau C 2012 Robustness against attacks of dual polarization encryption using the Stokes–Mueller formalism *J. Opt.* **14** 094004

[16] Li X, Lan T-H, Tien C-H and Gu M 2012 Three-dimensional orientation-unlimited polarization encryption by a single optically configured vectorial beam *Nature Commun.* **998** 1–6

[17] Lin C, Shen X, Hua B and Wang Z 2015 Three-dimensional polarization marked multiple-QR code encryption by optimizing a single vectorial beam *Opt. Commun.* **352** 25–32

[18] Zhu N, Wang Y, Liu J, Xie J and Zhang H 2009 Optical image encryption based on interference of polarized light *Opt. Express* **17** 13418–24

[19] Rajput S K and Nishchal N K 2012 Asymmetric color cryptosystem using polarization selective diffractive optical element and structured phase mask *Appl. Opt.* **51** 5377–86

[20] Rajput S K and Nishchal N K 2013 Image encryption using polarized light encoding and amplitude- and phase-truncation in Fresnel domain *Appl. Opt.* **52** 4343–52

[21] Rajput S K, Kumar D and Nishchal N K 2014 Photon counting imaging and polarized light encoding for secure image verification and hologram watermarking *J. Opt.* **16** 125406

[22] Lin C, Shen X, Wang Z and Zhao C 2014 Optical asymmetric cryptography based on elliptical polarized light linear truncation and a numerical reconstruction technique *Appl. Opt.* **53** 3920–8

[23] Maluenda D, Carnicer A, Martinez-Herrero R, Juvells I and Javidi B 2015 Optical encryption using photon-counting polarimetric imaging *Opt. Express* **23** 655–66

[24] Carnicer A, Juvells I, Javidi B and Martinez-Herrero R 2017 Optical encryption in the axial domain using beams with arbitrary polarization *Opt. Lasers Eng.* **89** 145–9

[25] Fatima A and Nishchal N K 2018 Optical image security using Stokes polarimetry of spatially variant polarized beam *Opt. Commun.* **417** 30–6

[26] Wang Q, Xiong D, Alfalou A and Brosseau C 2018 Optical image encryption method based on incoherent imaging and polarized light encoding *Opt. Commun.* **415** 56–63

[27] Zhang S and Karim M A 1999 Color image encryption using double random phase encoding *Microw. Opt. Technol. Lett.* **21** 318–23

[28] Joshi M, Shakher C and Singh K 2007 Color image encryption and decryption using fractional Fourier transform *Opt. Commun.* **279** 35–42

[29] Chen W and X Chen X 2011 Optical color image encryption based on asymmetric cryptosystem in the Fresnel domain *Opt. Commun.* **284** 3913–7

[30] Zhou N, Wang Y, Gong L, He H and Wu J 2011 Novel single channel color image encryption based on chaos and fractional Fourier transform *Opt. Commun.* **284** 2789–96

[31] Deng X and Zhao D 2012 Single channel color image encryption based on asymmetric cryptosystem *Opt. Laser Technol.* **44** 136–40

[32] Wei D, Ran Q, Li Y, Ma J and Tan L 2009 A convolution and product theorem for the linear canonical transform *IEEE Trans. Signal Process. Lett.* **16** 853–6

IOP Publishing

Optical Cryptosystems

Naveen K Nishchal

Chapter 7

Digital holography-based security schemes

7.1 Introduction

Holography is a successful technology for recording and reconstructing 3D objects/scenes. Earlier, this two-step lensless imaging process was called *wavefront reconstruction*. The two steps consist of two distinct operations: a recording step and a reconstruction step. It is a process of recording and displaying images based on the principle of optical interference and diffraction with or without the use of a lens. The interference between the waves reflected or transmitted by the object to be imaged and a reference wave creates the hologram [1].

Assuming $f(x,y)$ and $r(x,y)$ represent the wavefront to be detected and the reference wave, respectively. $\varphi(x,y)$ and $\psi(x,y)$ denote the phase values.

$$f(x, y) = |f(x, y) \exp[i\varphi(x, y)]| \tag{7.1}$$

$$r(x, y) = |r(x, y)\exp[i\psi(x, y)]| \tag{7.2}$$

These two waves interfere and the intensity of the sum can be expressed as

$$\begin{aligned} I(x, y) &= |f(x, y) \exp[i\varphi(x, y)] + r(x, y)\exp[i\psi(x, y)]|^2 \\ &= |f(x, y)|^2 + |r(x, y)|^2 \\ &\quad + 2|f(x, y)||r(x, y)|\cos[\psi(x, y) - \varphi(x, y)] \end{aligned} \tag{7.3}$$

The information of amplitude and the phase of the wavefront to be reconstructed is recorded and expressed by equation (7.3). The first two terms in the expression depend on the intensities of the individual waves and the third term depends on their relative phases. A recording medium is required for recording this intensity term, which is the interference pattern. The incident intensity must be linearly mapped onto the recording material. For reconstruction of the wavefront, $f(x,y)$, a coherent reconstruction wave illuminates the hologram.

doi:10.1088/978-0-7503-2220-1ch7

For the recording of a hologram, a coherent light source and a photographic film as a recording medium are required. A photographic film cannot be processed in real time due to the involved complexity, such as the wet chemical procedure. Recording a hologram is also considered as *sketching* of interference patterns, which, in principle, can be calculated using a computer and plotted on a transparency. The object that existed in the computer memory would be regenerated on illumination with the calculated reference wave. The holograms generated following this technique is called computer-generated holograms (CGH) and the process is defined as computer-generated holography. The holographic image can be seen with the help of a holographic 3D display that ignores the need of fabricating interference fringes each time. Any desired complex amplitude distribution can be designed using CGH, if it does not contradict the physical principles and technological limitations.

With the dawn of the digital era in the 1990s, the necessity to speed up the process of reconstruction of holograms was felt, which led to the development of a new subject, called digital holography [2, 3]. The recording of holograms digitally using a discrete electronic device such as CCD camera is referred to as *digital holography* (DH). The recorded hologram is termed as a digital hologram. The optical interference fringes recorded by the CCD camera are digitized into a 2D digital signal. The reconstruction is performed by simulating the propagation of the wavefield back to the plane of the object. The Huygens–Fresnel principle is the key concept in the numerical computation of light field propagation. The digital hologram is modulated with a replica of the original reference wave, in terms of wavelength and phase. Mathematically, it is expressed as the multiplication of the digital hologram with the complex amplitude of the reference wave. The interference fringe pattern (hologram) acquired electronically is digitized into a 2D digital signal and then processed using digital signal processing.

With the advent of digital holography and continuing advancement in megapixel CCD sensors having sufficient dynamic range in each pixel, digital holograms of useful size and in an appropriate form for numerical processing can be captured. Therefore, the requirement of time-consuming photographic process in optical holography is completely eliminated and it can now be performed in real time. The phase of the object wave, which cannot be measured directly but it could be recreated optically, therefore DH is becoming attractive for various applications. The storage of the digital hologram in a computer enables one to reduce the noise through image processing algorithms and numerically reconstruct the object with arbitrary views. Therefore, using DH fully complex information is stored and transmitted digitally. The information contained in the digital hologram can also be reconstructed optically after transmission by using SLMs. Since arbitrary views of 3D objects/scenes can be reconstructed digitally or optically, the method is applied in several applications including wavefront sensing, metrology, and holographic interferometry [3].

The digital holograms are complex-valued, hence, the size of the holograms are very large, which pose limitations while transmitting through conventional communication channels. To solve the issue, various data compression techniques have been developed, which compress a digital hologram for providing ease in digital transmission. The compression is achieved through applying quantization to the complex-valued holographic pixels. Digital hologram compression differs from normal image compression techniques because holograms contain 3D information in complex-valued pixels and inherent speckle noise. The speckle content gives the digital holograms a white-noise-like appearance. Depending upon the applications, lossy or lossless compression techniques can be applied for digital hologram compression.

There are several DH methods reported in literature, such as in-line DH, off-axis DH and phase-shifting DH. Usually Mach–Zehnder interferometer (MZI) or Michelson interferometer setups are used for recording the digital holograms. The recoding plane could be image plane, Fourier plane, Fresnel plane, fractional Fourier plane or gyrator plane.

The main problem with the DH is that upon numerical reconstruction the twin images and strong zero-order (dc) simultaneously appear in the output. Thus, the reconstruction image area contains the conjugate virtual images and the directly transmitted beam (zero-order or dc), which utilize the CCD space-bandwidth resources and therefore, result in a loss of resolution in the real image. In case of in-line DH, the twin images and zeroth-order overlap. Off-axis DH geometry is an effective configuration for separating the zero-order and the twin images. However, it is difficult to record a high quality off-axis DH due to the limitation of narrow bandwidth of CCDs. The zeroth-order term is very strong in case of Fourier hologram. The intensity of zero-order is reduced in the case of Fresnel holograms, which is further reduced when fractional order Fourier holograms are recorded. There are various techniques reported in the literature for the removal of zero-order and one of the undesired twin images. The phase-shifting interferometer technique solves both the issues. Amongst the various DH methods, digital phase-shifting interferometry stands out as a versatile and efficient method [3].

For good quality image reconstruction, DH requires a sensor with very high resolution. The CCD or CMOS camera to be used for recording holograms should have a large number of pixels and the pixel pitch should be very small.

7.2 Phase-shifting interferometry

Like conventional holography, DH consists of two processes: recording and reconstruction. In DH, photographic film is replaced by a CCD/CMOS camera for the process of recording the hologram. The image captured by the camera is transferred to the computer, which is digitally saved as a digital hologram in the computer memory. During the reconstruction process, this digitally stored hologram is accessed and reconstructed numerically with the help of virtual reference wave. The simulated reference wave similar to the one used during the process of recording, is employed for reconstruction of the object wavefront. Generally, a plane or a spherical wave is used for the recording purpose because these virtual reference waves can be generated easily and accurately in the computer. The objects are either a 3D body with diffusely reflecting surfaces or transparent, which is kept at some distance from the CCD/CMOS camera. Since the optical light field is in digital form, it gives the liberty to apply the techniques of digital signal processing. The speed of the reconstruction procedure depends only on the computer processor and efficiency of the reconstruction algorithm. There is a technique to find the Fresnel diffracted object wave at the plane of the recording medium. To reconstruct the complex object without actually calculating the complicated Fresnel transform, different intensity patterns are recorded by changing the phase of the reference wave by a known quantity.

Figure 7.1 shows the schematic of a MZI setup used for recording phase-shifting DH in the Fresnel plane. A polarized laser beam is divided into two beams by a beam splitter. One of the beams illuminates the 3D object and gets diffracted. The diffraction pattern propagates to the CCD camera located at some distance d. The paraxial region is considered so that scalar diffraction theory remains valid. It is for this reason the size of the illuminating object used is small enough compared to the CCD size. The diffracted beam interferes with a reference beam passed through two retarders, RP_1 and RP_2. Phase retardations in the reference beam of 0, $\pi/2$, π, and $3\pi/2$ with the direction of polarization of the incident light are achieved by controlling the positions of the fast and slow axes of the two plates. For introducing different phase-shifts, in addition to the use of retarders, piezoelectric transducers, liquid crystal variable retarders, liquid crystal SLMs, an acousto-optic modulator, and a tilted slab have been used. Phase-shifting can be applied to any holographic recording geometry.

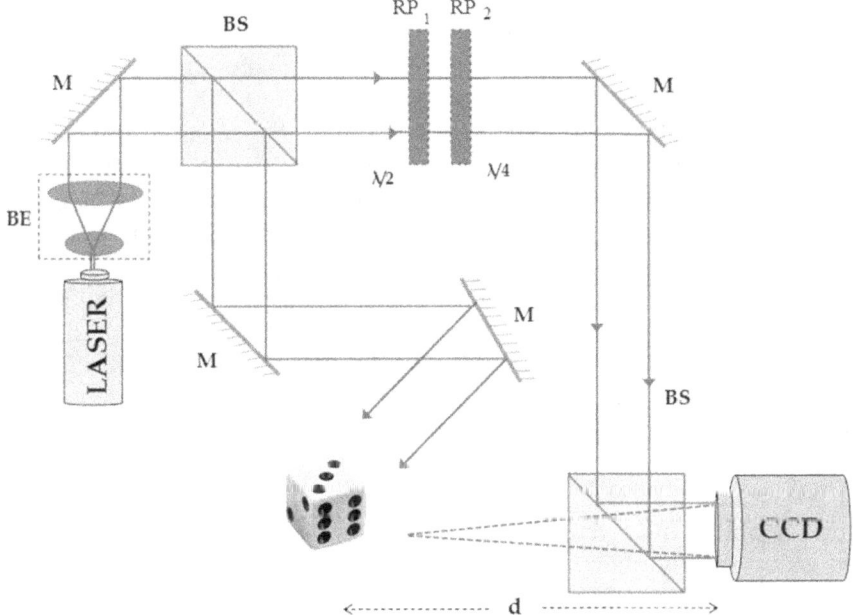

Figure 7.1. Optical setup for the digital hologram capture of real world objects. BE: beam expander, BS: beam splitter, RP: retardation plate, M: mirror, CCD: charge-coupled device camera.

In phase-shifting DH, multiple DHs are recorded corresponding to various phase differences between the object and the reference beam. There are different schemes of phase-shifting DH: four-step phase-shifting DH, three-step phase-shifting DH, and two-step phase-shifting DH. Consider the following expression representing the interference pattern:

$$I(x, y) = |f(x, y)|^2 + |r(x, y)|^2 + 2|f(x, y)||r(x, y)|\cos[\psi(x, y) - \varphi(x, y)] \quad (7.4)$$

In four-step phase-shifting DH, four DHs are recorded sequentially. The phase differences for the four DHs are 0, $\pi/2$, π, and $3\pi/2$. Therefore, on substituting 0, $\pi/2$, π, and $3\pi/2$ for $\psi(x,y)$, the obtained interference intensities are $I(x,y; 0)$, $I(x,y; \pi/2)$, $I(x,y; \pi)$, and $I(x,y; 3\pi/2)$, respectively. $\varphi(x,y)$ may be conveniently calculated using the following relation.

$$\varphi(x, y) = \arctan\left(\frac{I(x, y; 3\pi/2) - I(x, y; \pi/2)}{I(x, y; 0) - I(x, y; \pi)}\right) \quad (7.5)$$

Since the phase is determined, the amplitude $f(x,y)$ may be directly measured by detecting the intensity $|f(x, y)|^2$ on the photo detector when the reference beam is blocked. The wave field corresponding to the original object may be determined using the Fresnel integral by back-propagating the known field, as amplitude and phase are known.

In three-step phase-shifting DH, only three DHs are recorded sequentially. The phase differences for the three DHs are 0, $\pi/2$, and π. Therefore, on substituting 0, $\pi/2$, and π for $\psi(x,y)$, the obtained interference intensities are $I(x,y; 0)$, $I(x,y; \pi/2)$, and $I(x,y; \pi)$, respectively. In two-step phase-shifting DH, only two DHs are recorded sequentially. The phase differences for the two DHs are 0 and $\pi/2$. Therefore, on substituting 0 and $\pi/2$ for $\psi(x,y)$, the obtained interference intensities are $I(x,y; 0)$ and $I(x,y; \pi/2)$, respectively.

It may be noted that the phase step is always $\pi/2$ in phase-shifting DH. It is because for a fixed tolerance of the phase-step, the retrieved object field will have minimum error when phase shift of $\pm\pi/2$ is applied. Tolerance may be due to the inaccuracy of the phase shifter, vibration or air turbulence. The phase-shifting DH has a disadvantage that it cannot be applied to dynamic objects since different phase shifts of the reference are required [2, 3]. There are other phase-shifting geometries reported in literature such as parallel phase-shifting DH and single-shot phase-shifting DH. In single-shot DH, the complex field was obtained from the recorded hologram by solving a constrained optimization problem [4].

The experimental setup is modified in different ways to achieve in-line DH, off-axis DH with or without phase-shifting. When no phase-shifting is required, the retardation plates are simply removed. When some angle has to be introduced between the object and reference wave, the beam splitter placed before the CCD camera is accordingly manipulated. Since CCDs have relatively low resolution in comparison to a traditional recording medium, such as silver halide, the allowable angle between an object and the reference beam should be just a few degrees. Therefore, the distance between the object and CCD should be about 1 meter even with small objects in case of off-axis DH geometry. If a digital Fourier hologram of the 3D object has to be recorded then the object is placed at the focal plane of a Fourier transforming lens. Similarly, a digital fractional order Fourier hologram can be recorded.

7.3 Numerical reconstruction of digital holograms

If a hologram can be thought of as an aperture, then the reconstructed images are formed as a result of diffraction of the reference wave through this aperture. According to Huygen's principle, every point of a wavefront acts as a source for secondary spherical waves. The wavefront at all other places are the coherent superposition of these secondary wavelets. Fresnel approximation of the Fresnel–Kirchhoff integral describes the diffraction pattern obtained at a distance of d along the propagation direction of the wave, quantitatively, when the reference wave passes through the aperture or hologram [3]:

$$U(\xi, \eta) = \frac{ie^{ikd}}{\lambda d} \int\int_{-\infty}^{\infty} u(x, y)\exp\left\{-i\frac{k}{2d}[(x - \xi)^2 + (y - \eta)^2]\right\}dxdy \qquad (7.6)$$

Here, (x,y) and (ξ,η) represent the hologram and reconstruction plane coordinates, respectively. The coordinates in Fresnel propagation are shown in figure 7.2.

The intensity and phase of the reconstructed optical wavefront are given by equations (7.7) and (7.8), respectively. The terms *Im* and *Re* denote the imaginary and real parts, respectively.

$$I(\xi, \eta) = |U(\xi, \eta)|^2 \qquad (7.7)$$

$$\varphi(\xi, \eta) = \tan^{-1}\left(\frac{Im\{U(\xi, \eta)\}}{Re\{U(\xi, \eta)\}}\right) \qquad (7.8)$$

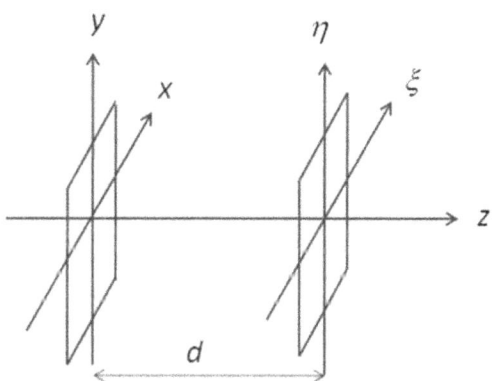

Hologram plane   Reconstruction plane

Figure 7.2. Coordinates in Fresnel propagation.

7.3.1 Discrete Fresnel transformation

Numerical reconstruction in DH may be done using the discrete Fresnel transformation approach. On expansion and rearrangement, equation (7.6) may be written as

$$
\begin{aligned}
\Gamma(\xi, \eta) = {} & \frac{i}{\lambda d} \exp\left(-i\frac{2\pi}{\lambda}d\right) \exp\left[-i\frac{\pi}{\lambda d}(\xi^2 + \eta^2)\right] \\
& \times \int_{-\infty}^{\infty}\int_{-\infty}^{\infty} U_r^*(x, y)U_h(x, y)\exp\left[-i\frac{\pi}{\lambda d}(x^2 + y^2)\right] \\
& \times \exp\left[i\frac{2\pi}{\lambda d}(x\xi + y\eta)\right]\mathrm{d}x\mathrm{d}y
\end{aligned}
\tag{7.9}
$$

For digital implementation, two substitutions are made in the Fresnel approximation.

$$
\nu = \frac{\xi}{\lambda d}; \ \mu = \frac{\eta}{\lambda d}
\tag{7.10}
$$

Suppose there is a raster of $N \times N$ points having a minimum step size of Δx and Δy along the coordinate axes (x,y). The pixel spacing of the CCD/CMOS device in the horizontal and vertical directions are Δx and Δy, respectively [3].

$$
\begin{aligned}
U_0(m\Delta x, n\Delta y) = {} & \frac{i}{\lambda d} \exp\left(-i\frac{2\pi}{\lambda}d\right) \exp[-i\pi\lambda d(m^2\Delta\nu^2 + n^2\Delta\mu^2)] \\
& \times \mathfrak{I}^{-1}\left\{U_0(k\Delta\xi, l\Delta\eta)\exp\left[-i\frac{\pi}{\lambda d}(k^2\Delta x^2 + l^2\Delta y^2)\right]\right\}
\end{aligned}
\tag{7.11}
$$

According to the communication theory, it is well known that for faithful reproduction of the original signal, the sampling rate must be greater than twice the bandwidth. Here, $\Delta\nu$ and $\Delta\mu$ are the sampling intervals in the observation plane, given as

$$
\Delta\nu = \frac{1}{N\Delta x}; \ \Delta\mu = \frac{1}{N\Delta y}
\tag{7.12}
$$

On substituting Δx, Δy, $\Delta\nu$ and $\Delta\mu$, Fresnel approximation is given by

$$
\begin{aligned}
\Gamma(m, n) = {} & \frac{i}{\lambda d} \exp\left(-i\frac{2\pi}{\lambda}d\right) \exp\left[-i\pi\lambda d\left(\frac{m^2}{N^2\Delta x^2} + \frac{n^2}{N^2\Delta y^2}\right)\right] \\
& \times \mathfrak{I}^{-1}\left\{U_r^*(k, l)U_h(k, l)\exp\left[-i\frac{\pi}{\lambda d}(k^2\Delta x^2 + l^2\Delta y^2)\right]\right\}
\end{aligned}
\tag{7.13}
$$

where $m = 0, 1,..., N - 1$; $n = 0, 1,..., N - 1$. Equation (7.13) is called the discrete Fresnel transform. For efficient computation, a fast Fourier transform (FFT) algorithm is used. For wavefront reconstruction, a sampling criterion from a carrier frequency signal as obtained in electronic holography has been reported [5].

7.3.2 Convolution approach

Numerical reconstruction in DH may also be done using the convolution approach. This method is based on the convolution theorem. Equation (7.14) is given as a superposition integral:

$$\Gamma(\xi, \eta) = \int_{-\infty}^{\infty} \int_{-\infty}^{\infty} U_h(x, y) U_r^*(x, y) g(\xi, \eta, x, y) \mathrm{d}x \mathrm{d}y \tag{7.14}$$

where the impulse response $g(\xi, \eta, x, y)$ is given as

$$g(\xi, \eta, x, y) = \frac{i}{\lambda} \frac{\exp\left[-i\frac{2\pi}{\lambda}\sqrt{d^2 + (x - \xi)^2 + (y - \eta)^2}\right]}{\sqrt{d^2 + (x - \xi)^2 + (y - \eta)^2}} \tag{7.15}$$

Equations (7.14) and (7.15) show that the linear system represented by $g(\xi, \eta, x, y) = g(\xi - x, \eta - y)$ is space-invariant. Therefore, the convolution theorem may be applied to equation (7.14). The convolution of two functions $f(x, y)$ and $g(x, y)$ is defined as

$$(f \otimes g)(x, y) = \int_{-\infty}^{\infty} \int_{-\infty}^{\infty} f(\xi, \eta) g(x - \xi, y - \eta) \mathrm{d}x \mathrm{d}y \tag{7.16}$$

where the convolution operation is denoted by $\otimes$. Applying convolution theorem,

$$\Gamma(\xi, \eta) = \mathfrak{I}^{-1}[\mathfrak{I}\{U_h(x, y). U_r^*(x, y)\}. \mathfrak{I}\{g(x, y)\}] \tag{7.17}$$

It can be clearly seen that equation (7.17) has two forward Fourier transformations and one inverse Fourier transformation. The coordinates can be shifted by $N/2$ in order to make the reconstructed area symmetrical with respect to the optical axis.

$$g(k, l) = \frac{i}{\lambda} \frac{\exp\left[-i\frac{2\pi}{\lambda}\sqrt{d^2 + \left(k - \frac{N}{2}\right)^2 \Delta x^2 + \left(l - \frac{N}{2}\right)^2 \Delta y^2}\right]}{\sqrt{d^2 + \left(k - \frac{N}{2}\right)^2 \Delta x^2 + \left(l - \frac{N}{2}\right)^2 \Delta y^2}} \tag{7.18}$$

The Fourier transform of $g(k, l)$ may be analytically expressed as in equation (7.17):

$$G(n, m) = \exp\left\{-i\frac{2\pi d}{\lambda}\sqrt{1 - \frac{\lambda^2\left(n + \frac{N^2\Delta x^2}{2d\lambda}\right)^2}{N^2\Delta x^2} - \frac{\lambda^2\left(m + \frac{N^2\Delta y^2}{2d\lambda}\right)}{N^2\Delta y^2}}\right\} \tag{7.19}$$

This saves one Fourier transform operation for the reconstruction of the real image, which in turn would save processing time:

$$\Gamma(\xi, \eta) = \mathfrak{I}^{-1}\{\mathfrak{I}(U_h. U_r^*). G\} \tag{7.20}$$

The Fresnel approximation and the convolution approach are the two basic methods of numerical reconstruction in DH. The FFT algorithm is used for implementing both methods. Only one forward FFT is performed in the Fresnel approximation, but two or three FFTs are performed in the case of the convolution approach. Therefore, the result of the Fresnel approximation is in the frequency domain due to the single Fourier transform and that of the convolution approach is still in the spatial domain because of one forward Fourier transform and an inverse Fourier transform. As a consequence of this difference, the pixel sizes of the reconstructed image, $\Delta\xi$ and $\Delta\eta$, depend on the reconstruction distance and the wavelength in the case of Fresnel approximation, which is given in equation (7.12). However, for the convolution approach, the pixel sizes of the reconstructed image as well as hologram become equal, and are hence independent of the wavelength and distance. Therefore, a CCD/CMOS device with a smaller pixel size can be used for achieving a higher resolution during reconstruction process. But, the actual resolution is governed by the physical parameters of the optical system.

Selection of the reconstruction method depends upon the requirements of the specific application. For an object or surface bigger than the CCD/CMOS sensor, generally, the Fresnel approximation is used because the algorithm takes less processing time. But for applications in particle measurement and microscopy to detect very small objects, the convolution approach is more suitable since it is more accurate.

7.3.3 Angular spectrum method

For numerical reconstruction using Fresnel approximation and the convolution approach, it is required that the object must be placed at a distance more than some minimum value. Therefore, these approaches cannot be used in all the cases. Both the methods have the same limitation, i.e. the spatial frequency of the detector is too low and aliasing takes place if the distance between object and detector is less than the minimum value, which is given as

$$d_{\min} = \frac{(N\Delta x)^2}{N\lambda} \tag{7.21}$$

Here, Δx and N denote size and number of the pixels, respectively. However, the angular spectrum method does not have this limitation. At the same time, it has higher accuracy and comparable computational efficiency in comparison to the other two methods.

Suppose the optical wavefield is $U_0(x,y;0)$ at the plane $d = 0$, equation (7.22) gives the angular spectrum in this plane, which is written as $A(k_x,k_y;0)$ by taking the Fourier transform:

$$A(k_x, k_y; 0) = \int\int U_0(x, y; 0)\exp\left[-i(k_x x + k_y y)\right]\mathrm{d}x\mathrm{d}y \tag{7.22}$$

where k_x and k_y are spatial frequencies along coordinates x and y, respectively. $A(k_x,k_y;d)$ represents the angular spectrum at distance d which is calculated from $A(k_x,k_y;0)$ and is given by

$$A(k_x, k_y; d) = A(k_x, k_y; 0)\exp(ik_z d) \tag{7.23}$$

where $k_z = \sqrt{k - k_x - k_y}$. The reconstructed complex wavefield at a plane perpendicular to the z-axis, along which the wavefield is propagating, is given by equation (7.24).

$$\begin{aligned} U(\xi, \eta; d) &= \int\int A(k_x, k_y; d)\exp\left[i(k_x\xi + k_y\eta)\right]dk_x dk_y \\ &= \Im^{-1}[\Im(U_0)\exp(ik_z d)] \end{aligned} \tag{7.24}$$

7.4 Information security using digital holography

Under the DRPE framework, the encrypted data is fully complex, which needs to be recorded and stored holographically. Information recorded in this way has difficulty in transmission over digital communication channels. Since DH is based on the digital acquisition of interference pattern, techniques based on DH have important benefits in the field of information security. They enable digital storage, transmission, and real-time decryption of encrypted data. The encryption is performed directly on the complex information and decryption does not require heavier computation than a usual image reconstruction procedure. Therefore, there is a potential advantage in terms of processing speed over the fully digital techniques. Also, there is no compromise with the strength of the security of the technique.

For the first time, in the year 2000, a DRPE scheme was combined with a digital holographic technique, in which an input image was encrypted optically and recorded as a digital hologram. The RPM was also recorded as a key digital hologram. For image retrieval, numerical reconstruction was applied to the encrypted digital hologram, which was modulated with the key digital hologram [6]. The encrypted hologram and key hologram can be transmitted over a communication channel and the original image can be retrieved electronically. To make full use of CCD space-bandwidth and reduce the object distance, phase-shifting interferometry has been proposed [7]. In this technique, only the real image is obtained and zero-order is suppressed. The method was extended for securing a 3D object employing phase-shifting interferometer architecture [8, 9]. The DH was recorded in the Fresnel diffraction region. A 3D object with different perspectives and focused at different planes can be reconstructed digitally or optically after decryption with the use of correct keys. The focusing at different planes is achieved with numerical reconstruction for different (positive or negative) Fresnel propagation distances. A DH-based security system is shown in figure 7.3. Combining DH with DRPE offers several advantages such as speed, large degrees of freedom, high security, and electronic acquisition and transmission.

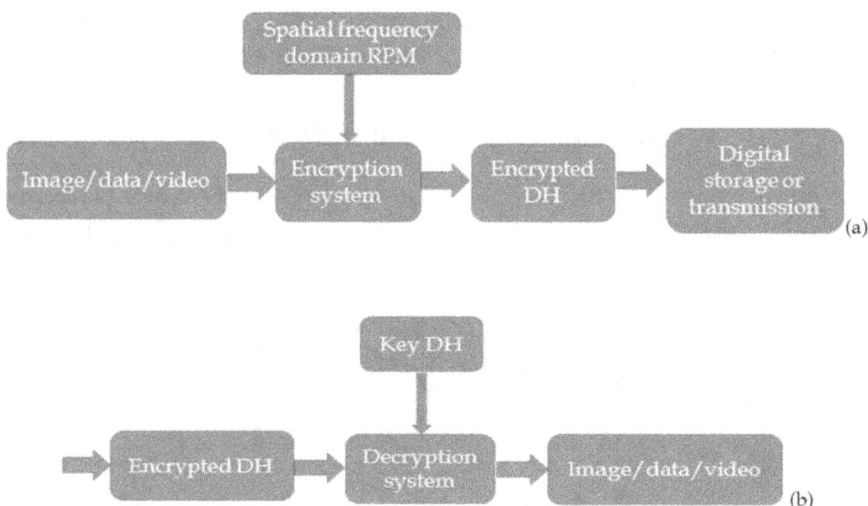

Figure 7.3. DH-based security system. (a) Data encoding into encrypted DH and transmission and (b) data receiver and decryption.

The liquid crystal SLMs can be operated in either phase-only mode or in amplitude mode, a digital hologram can be reconstructed optically with phase-only information. Utilizing an SLM, a real-time 3D object reconstruction has been reported [10]. This was achieved with a phase-encoded digital hologram. The use of phase-only information enabled reduction in the data and optical reconstruction without much loss of optical power. In order to reduce the size of the digital hologram for ease in transmission, data compression techniques have been applied to encrypted 3D objects. Lossy quantization was combined with lossless coding techniques to quantify compression ratios [11–14]. For data compression, the digital holograms were stored in native MATLAB floating-point representation with 8 bytes of real information and 8 bytes of imaginary information for each pixel. A parameter, compression ratio is used for the performance measure, which is defined as the ratio of uncompressed size to the compressed size. Lossless techniques perform very poorly on an encrypted digital hologram due to white noise characteristics. Quantization has very good effect and reductions to as few as 3 bits in each real and imaginary part have resulted in a good decompressed and decrypted 3D object [13].

A fully digital technique using DRPE in Fourier and FRT domain encoding has been reported, in which the input image multiplied with an RPM was Fourier or fractional Fourier transformed. Using interference with a wave from another RPM, the encrypted data was captured using a CCD camera. The decryption key was also recorded as a digital hologram. To further enhance the security, an electronic key was multiplied with the encrypted hologram and a Fourier or FRT was obtained. Numerical reconstruction was carried out using the electronic key and the key hologram [15]. The study was further extended for fully-phase encryption [16]. In this method, the input amplitude image is phase-encoded and then its Fourier or

Fresnel digital hologram is recorded using RPMs. The decrypted image obtained after using correct keys is a phase image, which needs to be converted into an amplitude image. Therefore, to obtain phase contrast a digital reference wave, R_D, which must be a replica of the experimental reference wave, is used [17].

$$R_D = A_R \exp\left[i\left(\frac{2\pi}{\lambda}\right)(k_x m \Delta x + k_y n \Delta y) \right] \qquad (7.25)$$

where A_R is the amplitude, whose adjustment is not of particular importance because its value is generally set to unity. Δx and Δy denote the pixel dimensions of the CCD/CMOS camera. The parameters k_x and k_y are the wave vector components that must be adjusted such that the propagation direction matches as closely as possible with the actual reference wave used in the experiment.

In chapter 5, watermarking of 3D objects through DH has been discussed. Such a study has been extended to the discrete-cosine transform domain also [18]. The watermark image was recorded as a digital hologram and then superposed onto the discrete-cosine transform coefficients of the content image. A review on 3D image encryption, transmission, and digital processing using DH has been reported [19]. In DH, numerical reconstruction of the hologram is possible with either phase-only or amplitude-only information. Reconstruction results based on the phase-only or amplitude-only have been compared, and it has been observed that phase-only information produces better quality. For improving the quality of reconstruction various digital filters such as mean filter, median filter are applied. Also, digital techniques have been applied for removing the speckles. A technique for binary image/data encryption using a phase-shifting interferometer has been proposed, which employed successive forward and backward encryption/transmission/decryption procedure over a digital communication channel [20].

A flexible optical encryption with multiple users and multiple security levels based on multiplexing has been reported. In the scheme, normal users can decrypt different private images/information from the same encrypted data but a superuser can have a key that can decrypt all encrypted information, and multiplexed images/data can be encrypted with different levels of security. The proposal was illustrated through a real-world 3D scene, captured with phase-shifting DH, and encrypted using FRT, where different users have access to different 3D objects in the scene [21]. The schematic of the encryption-decryption system is shown in figure 7.4. An input 3D object/scene was segmented into different parts and each segment was encrypted using FRT domain DRPE. The 3D object was recorded and reconstructed from a phase-shifting Fresnel digital hologram.

In DII experiments, the limitations of the real optical elements (diameter of the lens and surface area) restricts the optical aperture, which actually limits the field-of-view and causes difficulty while encrypting large-sized images. Therefore, ptychographic scanning operation has been introduced into DH with the idea of expanding the field-of-view [22]. The original image/data was translated sequentially for ptychographic scanning and corresponding pairs of phase-shifted interferograms were recorded as ciphertexts. For retrieval, holographic processing and ptychographic

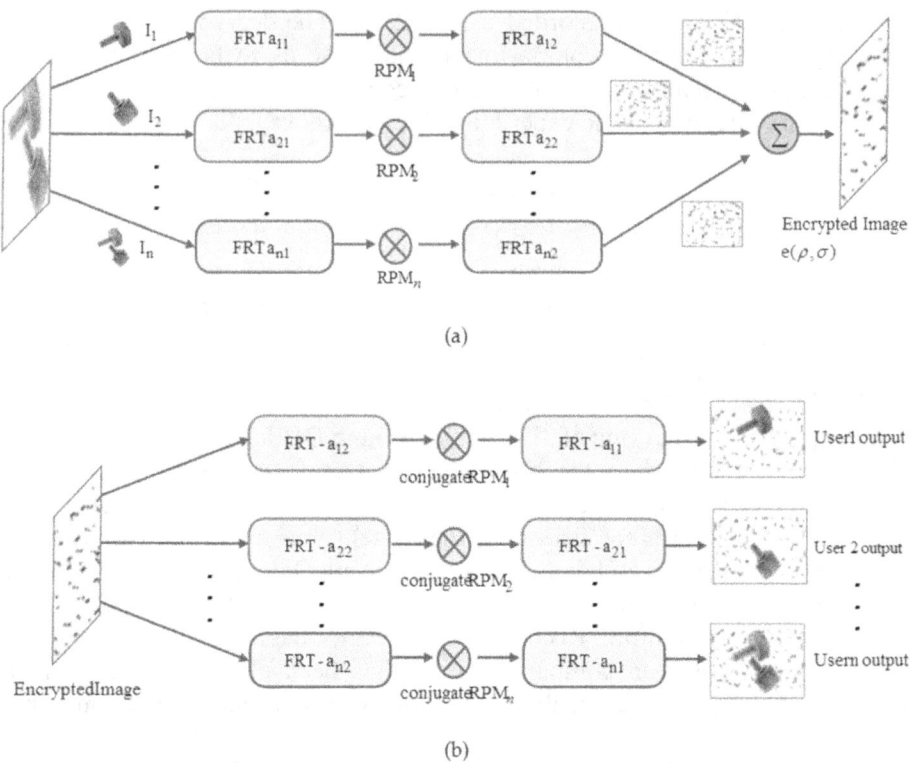

(a)

(b)

Figure 7.4. Schematic of the (a) encryption system and (b) decryption system. FRT: fractional Fourier transform, RPM: random phase mask. Reprinted from [21], Copyright (2011), with permission from Elsevier.

iterative reconstruction were applied. Ptychography is a coherent imaging method that uses a movable aperture to scan a specimen and reconstruction is carried out by the sequentially recorded diffraction patterns.

An optical security system based on polarization DH and a QR code with four-dimensional key design in amplitude, phase, polarization, and propagation distance has been demonstrated. In this study, a single-shot off-axis DH with orthogonal polarized reference waves was used to perform the polarization state recording. Two polarization holograms were calculated and fabricated and were used in the reference arms for generating random amplitude and phase distribution, respectively. For reconstruction, original information represented with a QR code was retrieved using Fresnel propagation and correct decryption keys [23]. Further, 3D encryption and anti-counterfeiting using DH has been reported [24]. In this method, arbitrary micro phase-step DH with optical interferometry and digital image processing were used for obtaining the information about a 3D object and keys. For holographic encryption, CGH was used.

A 3D object encryption using diffractive imaging and DH has been demonstrated. In this study, a digital hologram of a microlens array was recorded. The microlens array was fabricated using the thermal reflow method. The hologram encoding was performed based on multiple intensity samplings of the complex-amplitude wavefront with axial translation of the CCD in the FRT domain. The function was then Fresnel propagated for three different locations of the sensor, and corresponding diffraction patterns were recorded. The diffraction patterns were referred to as cipher-texts. For decryption, an iterative phase retrieval algorithm was applied to extract the hologram from corresponding encrypted images. The corresponding phase profile of the microlens array was then obtained [25]. An image encryption method based on phase mask multiplexing and photon counting imaging for multiple image authentications and digital hologram security has also been reported. In this scheme, multiple images to be authenticated were converted into phase-only functions using a phase retrieval algorithm in the FRT domain, which were then multiplexed into a single phase-only function and further encrypted into a complex-valued function. For decryption, the photon-limited decrypted images were obtained after applying the appropriate keys. Such images contained sufficient information for verification because of sparse representation [26]. Recently, DH has been used in securing voice, in which the optical voice was recorded as a hologram and encrypted using DRPE [27]. The recording was done with a high-speed image sensor. The optical voice encryption concept can be applied further in voice identification and watermarking.

7.5 Digital holography-based geometries for image encryption

In the previous sections, the basics of DH recording and numerical reconstruction for Fresnel DH have been explained. In the following sub-sections, two methods of image encryption through DH, Fourier domain and fractional Fourier domain encoding are discussed.

7.5.1 Fourier domain DRPE through digital holography

In this section, image encryption using double random Fourier domain encoding through DH is discussed [15]. Let real-valued function $f(x,y)$ denote the original 2D image to be encrypted, and (x,y) and (u,v) denote the input and Fourier domain coordinates, respectively. The original image is multiplied by an RPM, $p(x,y)$ [= exp$\{i2\pi R_1(x,y)\}$] and its Fourier transformation is obtained, which interferes with a wave from another RPM, $q(u,v)$ [= exp$\{i2\pi R_2(u,v)\}$] and the encrypted interference pattern is recorded by a CCD/CMOS camera. The intensity $I_H(u,v)$ of the digital hologram obtained by the interference between these two waves in the Fourier domain is expressed as

$$
\begin{aligned}
I_H(u, v) = &|\{F(u, v) \otimes P(u, v)\} + q(u, v)|^2 \\
&|\{F(u, v) \otimes P(u, v)\}|^2 + |q(u, v)|^2 + \\
&\{F(u, v) \otimes P(u, v)\}^* q(u, v) \\
&+ \{F(u, v) \otimes P(u, v)\} q^*(u, v)
\end{aligned}
\tag{7.26}
$$

where $F(u,v)$ and $P(u,v)$ denote the Fourier transform of $f(x,y)$ and $p(x,y)$, respectively. The information about the first and second terms on the right-hand side of equation (7.26) can be obtained *a priori* by capturing the power spectrum of the encrypted data and the reference beam. After subtracting the captured intensity distributions, the holographic data remains as

$$I_H(u, v) = \{F(u, v) \otimes P(u, v)\}^* q(u, v) + \{F(u, v) \otimes P(u, v)\} q^*(u, v) \quad (7.27)$$

For capturing the key digital hologram, the original object, input plane RPM and the transforming lens are removed from the experimental setup. A plane wave of uniform amplitude then illuminates the RPM.

$$\begin{aligned} I_K(u, v) &= |1 + q(u, v)|^2 \\ &= |1|^2 + |q(u, v)|^2 + q^*(u, v) + q(u, v) \end{aligned} \quad (7.28)$$

Applying a process similar to equation (7.27), the expression for key holographic data can be written as,

$$I_K(u, v) = q^*(u, v) + q(u, v) \quad (7.29)$$

For extracting the holographic data from equations (7.27) and (7.29), retaining the 4th term in both the expressions they are multiplied together. The function $\{F(u, v) \otimes P(u, v)\}$ is then obtained, which upon inverse Fourier transformation gives the original real-valued function, $f(x,y)$. The digital signal processing algorithms are applied suitably to the holograms.

$$\{F(u, v) \otimes P(u, v)\} \times q^*(u, v) \times q^*(u, v) \Rightarrow \{F(u, v) \otimes P(u, v)\}$$
$$\mathfrak{F}^{-1}[\{F(u, v) \otimes P(u, v)\}] = |f(x, y)p(x, y)|^2 = f(x, y) \quad (7.30)$$

The experimental results based on the above-discussed principle of Fourier domain DRPE and DH have been shown in figure 7.5.

7.5.2 FRT domain DRPE through digital holography

In this section, image encryption using FRT domain encoding through DH is discussed [15]. FRT has been used to enhance the key size and hence enhanced level of security. The symbols used for different parameters have been taken from chapter 2 and the previous section. The real-valued function is multiplied by an RPM and its FRT of order α is performed.

$$g(u, v) = \mathfrak{F}^{\alpha}[f(x, y)p(x, y)] \quad (7.31)$$

(a) (b) (c) (d)

Figure 7.5. Experimental results for Fourier domain DH. (a) Original image captured using a CCD camera, (b) captured digital encrypted Fourier hologram, (c) captured digital key hologram, and (d) decrypted image obtained after numerical reconstruction and using the key hologram. Reprinted from [15], Copyright (2004), with permission from Elsevier.

The FRT parameter, $\alpha = a\pi/2$. A lens performs ath order FRT if the distance d and the focal length of the lens f satisfy the following relation. Please refer to figure 2.3.

$$d = f_s \tan(\alpha/2)$$
$$f = f_s/\sin \alpha \tag{7.32}$$

Here, f_s is a scale factor. Using interference with a wave from another RPM, q (u,v), the encrypted hologram is recorded using a CCD/CMOS camera. The intensity $I_H(u,v)$ of the digital hologram obtained by the interference between these two waves in the FRT domain is expressed as

$$\begin{aligned}I_H(u, v) &= |g(u, v) + q(u, v)|^2 \\ &= |g(u, v)|^2 + |q(u, v)|^2 + g^*(u, v)q(u, v) + g(u, v)q^*(u, v)\end{aligned} \tag{7.33}$$

Following the procedures explained in the previous section, the holographic data can be expressed as

$$I_H(u, v) - g^*(u, v)q(u, v) + g(u, v)q^*(u, v) \tag{7.34}$$

A key DH is also captured as defined in equations (7.28) and (7.29). For numerical reconstruction and the retrieval of the original function from the encrypted DH, the procedures explained in equation (7.30) have to be repeated. The major difference is that instead of Fourier transform, FRT of the same order, which was used during encryption, has to be performed.

Various optical geometries for DH-based security techniques have been discussed. The main idea is to enhance the key space and hence make the cryptosystem very powerful. One of such techniques is encoding the input amplitude image/data into fully-phase mode. Fully-phase encryption using DH has been shown to be stronger as compared to amplitude encoding [16]. In conclusion, it can be stated that DH has the potential for a powerful practical cryptosystem.

References

[1] Hariharan P 2002 *Basics of Holography* (Cambridge: Cambridge University Press)

[2] Poon T-C and Liu J-P 2014 *Introduction to Modern Digital Holography with MATLAB* (Cambridge: Cambridge University Press)

[3] Schnars U, Falldorf C, Watson J and Juptner W 2015 *Digital Holography and Wavefront Sensing; Principles, Techniques, and Applications* (Berlin: Springer)

[4] Khare K, Ali P T S and Joseph J 2013 Single shot high resolution digital holography *Opt. Express* **21** 2581–91

[5] Khare K and George N 2003 Direct coarse sampling of electronic holograms *Opt. Lett.* **28** 1004–6

[6] Javidi B and Nomura T 2000 Securing information by use of digital holography *Opt. Lett.* **25** 28–30

[7] Lai S and Neifeld M A 2000 Digital wavefront reconstruction and its application to image encryption *Opt. Commun.* **178** 283–9

[8] Tajahuerce E and Javidi B 2000 Encrypting three dimensional information with digital holography *Appl. Opt.* **39** 6595–601

[9] Tajahuerce E and Javidi B 2001 Three-dimensional image security *Proc. SPIE* **10298** 102980G

[10] Matoba O, Naughton T J, Frauel Y, Bertaux N and Javidi B 2002 Real-time three-dimensional object reconstruction by use of a phase-encoded digital holograms *Appl. Opt.* **41** 618792

[11] Nomura T, Okazaki A, Kameda M, Morimoto Y and Javidi B 2001 Digital holographic data reconstruction with data compression *Proc. SPIE* **4471** 235–42

[12] Naughton T J, Tajahuerce E, McDonald J B and Javidi B 2004 Three-dimensional image sensing, encryption, compression, and transmission using digital holography *Proc. SPIE* **5611** 24–32

[13] Naughton T J and Javidi B 2004 Compression of encrypted three-dimensional objects using digital holography *Opt. Eng.* **43** 2233–8

[14] Pitkaaho T, Pitkakangas V, Niemela M, Rajput S K, Nishchal N K and Naughton T J 2018 Space-variant video compression and processing in digital holographic microscopy sensor networks with application to potable water monitoring *Appl. Opt.* **57** E190–8

[15] Nishchal N K, Joseph J and Singh K 2004 Securing information using fractional Fourier transform in digital holography *Opt. Commun.* **235** 253–9

[16] Nishchal N K, Joseph J and Singh K 2004 Fully phase encryption using digital holography *Opt. Eng.* **43** 2959–66

[17] Cuche E, Marquet P and Depeursinge C 1999 Simultaneous amplitude-contrast and quantitative phase-contrast microscopy by numerical reconstruction of Fresnel off-axis holograms *Appl. Opt.* **38** 6994–7001

[18] Chang H T and Tsan C L 2005 Image watermarking by use of digital holography embedded in the discrete-cosine-transform domain *Appl. Opt.* **44** 6211–9

[19] Javidi B, Tajahuerce E, Naughton T J, Frauel Y and Matoba O 2005 Three-dimensional image encryption, transmission and processing by using digital holography *Proc. SPIE* **5954** 595403

[20] Gil S K, Jeon S H, Kim N and Jeong J R 2006 Successive encryption and transmission with phase-shifting digital holography *Proc. SPIE* **6136** 613615

[21] Nishchal N K and Naughton T J 2011 Flexible optical encryption with multiple users and multiple security levels *Opt. Commun.* **284** 735–9

[22] Gao Q, Wang Y, Li T and Shi Y 2014 Optical encryption of unlimited-size images based on ptychographic scanning digital holography *Appl. Opt.* **53** 4700–7

[23] Lin C, Shen X and Li B 2014 Four-dimensional key design in amplitude, phase, polarization and distance for optical encryption based on polarization digital holography and QR code *Opt. Express* **22** 20727–39

[24] Shiu M T, Chew Y K, Chan H T, Wong X Y and Chang C C 2015 Three-dimensional information encryption and anticounterfeiting using digital holography *Appl. Opt.* **54** A84–8

[25] Mehra I, Singh K, Agarwal A K, Gopinathan U and Nishchal N K 2015 Encrypting digital hologram of three-dimensional object using diffractive imaging *J. Opt.* **17** 035707

[26] Rajput S K, Kumar D and Nishchal N K 2015 Optical encryption system based on phase mask multiplexing and photon counting imaging for multiple image authentication and digital hologram security *Appl. Opt.* **54** 1657–66

[27] Rajput S K and Matoba O 2017 Optical voice encryption based on digital holography *Opt. Lett.* **42** 4619–22

IOP Publishing

Optical Cryptosystems

Naveen K Nishchal

Chapter 8

Securing fused multispectral data

8.1 Introduction

The term fusion in general is defined as an approach for extracting information acquired from various domains. Image fusion is very popular because it provides improved image quality after the fusion of extracted information from different sources. A fused image offers better interpretation/analysis of image data. Fusion techniques play a significant role in multi-wavelength phenomena. Therefore, the applications of image fusion are diverse, such as remote sensing, astronomy, security and surveillance systems, agriculture, and medical imaging. The term quality of the fused image and its computation technique depend upon the specific application [1]. The fusion methods can be broadly classified into spatial domain fusion and frequency domain fusion. Most of the digital schemes use input domain fusion while optical techniques are mainly based on frequency domain fusion.

For the same 3D scene and modality, which may be blurred and noisy, fusing two or more images leads to a deblurred and denoised image. In superresolution fusion, blurred input images with low spatial resolution are fused together, which provides a high-resolution image. Imaging devices having low resolution include smart mobile phones, web cameras, camcorders, security and surveillance cameras. To achieve digital image fusion, various approaches are followed [2].

- **Multi-sensor approach**: to achieve high spatial and spectral resolutions. This is performed by combining images from different sensors, e.g. fusion of images captured by a visible and infrared camera. One of the sensors may have high spatial resolution and the other one high spectral resolution. A selection of sensor types depends on the need or application. Of late, drone-based multi-spectral sensing has been in use. It helps the decision-making process in various applications where physical access or reach is difficult.
- **Multi-view approach**: using the same sensor, a set of images of the same scene is captured. Different viewpoints of the same scene are fused, which yield an image with higher resolution.

doi:10.1088/978-0-7503-2220-1ch8

- **Multi-temporal approach**: at different times, images of the same scene can be captured. This will enable the evaluating of changes in the scene or obtaining a less degraded image of the scene. The approach can also be used in medical imaging where change in organs and tumors can be detected. Also, in remote sensing for monitoring land or forest exploitation. In these applications, the image acquisition period may be months or years depending upon the requirement.
- **Multi-focus approach**: for capturing a 3D scene, a lens system with varying focal lengths can be used repeatedly.

The goal of the image fusion is to integrate the information obtained from one or a combination of the above four approaches and then transform into a fused image that contains improved information quality that cannot be achieved otherwise. In other words, image fusion combines higher spatial information with high spectral information to create a synthetic high-resolution image. *Image registration* and *combining the image functions* such as intensity, color, etc, are the two stages of digital image fusion. Image registration involves four operations: feature detection, feature matching, transform model estimation, and resampling of the image.

Considering the diversity of applications, it is not possible to design a universally acceptable fusion method. Therefore, a fusion method should consider not only the fusion purpose but also the characteristics of individual sensors used for image capture. Also, imaging conditions, geometry, effects of noise, accuracy, and data properties should be taken into account. Therefore, a trade-off is always needed for achieving an acceptable quality of a fused image [2].

There are three levels of abstraction in the digital image fusion process. They are pixel level, feature level, and decision level fusion. Figure 8.1 shows the algorithms

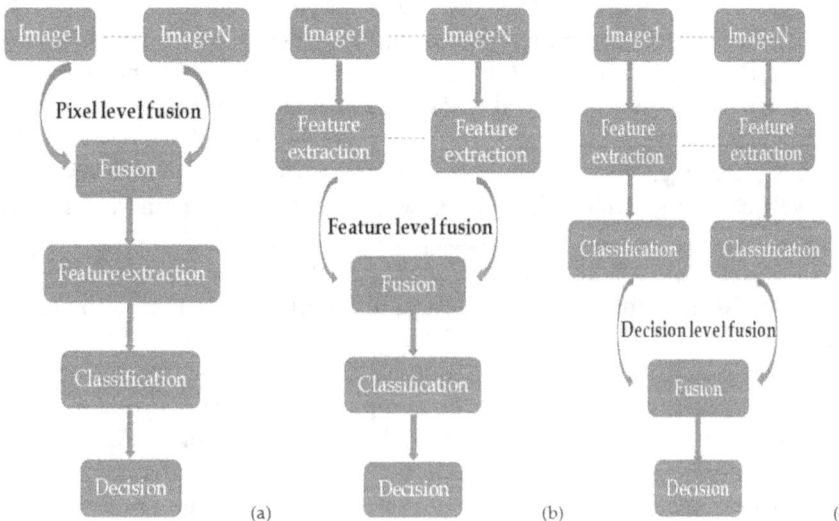

Figure 8.1. Different digital image fusion levels. (a) Pixel level fusion, (b) feature level fusion, and (c) decision level fusion.

for different levels of digital image fusion. In *pixel level fusion*, a number of images are fused together and then feature extraction, classification, and decision-making steps are followed. In *feature level fusion*, features are extracted from each image first and then the fusion process takes place. Further, classification and decision-making steps are followed. In *decision level fusion*, features are extracted from each image. With extracted features, images are classified and fused together. Further, the decision-step is followed.

Another important point that needs attention is that there are two terminologies associated with image fusion. They are hyperspectral images and multispectral images. Hyperspectral imaging integrates images collected across the electromagnetic spectrum. Usually, a hyperspectral image could have hundreds or even thousands of much narrower bands with ranges maybe 10–20 nm. The hyperspectral image is a 3D datacube consisting of 2D spatial and 1D spectral information. Multispectral imaging integrates images captured within specific wavelength range across the electromagnetic spectrum. It refers usually three to ten bands. In military applications, usually targets are measured with mid- and long-infrared wavelengths. This is also referred to as multispectral imaging. In surveillance systems, multiple spectral bands such as millimeter wave radar, terahertz camera, infrared camera are used to detect threats [3]. Therefore, fusion becomes very important for enabling early detection of occurring threats. Multispectral detection is a basis of all security systems.

8.2 Image fusion principle using wavelet transform

In this section, the principle of image fusion using wavelet transform (WT) is discussed. Fusion of a color image after decomposing into primary color components has been considered as an example. WT is one of the image processing tools, which has attractive capabilities such as multi-resolution, denoising, and feature extraction [4]. A 1D mother wavelet $h(x)$ is a finite-duration window function, which can generate a family of daughter wavelets by varying dilation, a, and shift, b. It is expressed as

$$h_{a,b}(x) = \frac{1}{\sqrt{a}} h\left(\frac{x-b}{a}\right) \tag{8.1}$$

The mother wavelet must satisfy the admissibility conditions. The conditions are: (i) it must be oscillatory, (ii) should have fast decay to zero, and (iii) integrate to zero. The WT decomposes a signal into four wavelet coefficients: approximation, vertical, horizontal, and diagonal coefficients. Figure 8.2 shows the decomposition process.

A 1D function $f(x)$ is decomposed into its wavelet coefficients with scale m_0 as [4]:

$$W_\varphi(m_0,k) = \left(\frac{1}{\sqrt{L}}\right) \sum_k f(x)\varphi_{m_0,k} \tag{8.2}$$

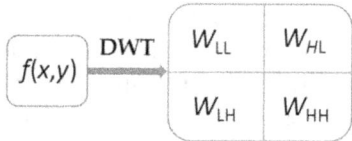

Figure 8.2. Decomposition of wavelet coefficients of an image. DWT: discrete WT.

$$W_\psi(m,k) = \left(\frac{1}{\sqrt{L}}\right)\sum_k f(x)\psi_{m,k} \tag{8.3}$$

Here, L denotes the scaling parameter of the WT, $W_\varphi(m_0,k)$ and $W_\psi(m,k)$ represent approximation and detailed coefficients, respectively. For performing 2D discrete WT, a 1D discrete WT is performed on the rows first and then on the columns. This results in a set of approximation coefficients $W_\varphi(m,r,s)$ and three sets of detailed coefficients, $W^\eta_\psi(m,r,s)$ where $\eta = \{H,V,D\}$ stands for horizontal, vertical, and diagonal components. A color image is separated into primary color components (red, green, and blue). The wavelengths corresponding to each color component are fused together using discrete WT to obtain a better quality color image.

Let $f_r(x,y)$, $f_g(x,y)$, and $f_b(x,y)$ denote red, green, and blue color components, respectively, of a color image, $f_c(x,y)$. Consider that the recovered images corresponding to wavelengths λ_r and $\lambda_r + \delta\lambda_r$ are given as $f_{r1}(x,y)$ and $f_{r2}(x,y)$, respectively. Similarly, the recovered images corresponding to wavelengths λ_g and $\lambda_g + \delta\lambda_g$ are given as $f_{g1}(x,y)$ and $f_{g2}(x,y)$, respectively. Similarly, the recovered images corresponding to wavelengths λ_b and $\lambda_b + \delta\lambda_b$ are given as $f_{b1}(x,y)$ and $f_{b2}(x,y)$, respectively. For achieving image fusion, discrete WT is applied.

$$W_{\mathrm{FLLr}} = \mathrm{AVG}(W_{\mathrm{LLr1}}, W_{\mathrm{LLr2}}) \tag{8.4}$$

$$W_{\mathrm{FHLr}} = \mathrm{AVG}(W_{\mathrm{HLr1}}, W_{\mathrm{HLr2}}) \tag{8.5}$$

$$W_{\mathrm{FLHr}} = \mathrm{AVG}(W_{\mathrm{LHr1}}, W_{\mathrm{LHr2}}) \tag{8.6}$$

$$W_{\mathrm{FHHr}} = \mathrm{AVG}(W_{\mathrm{HHr1}}, W_{\mathrm{HHr2}}) \tag{8.7}$$

The combination of all four components can be expressed as

$$g_R(\psi,\varphi) = \left\{ W_{FLL_R}, W_{FHL_R}, W_{FLH_R}, W_{FHH_R}, \right\} \tag{8.8}$$

Similarly for green and blue color components, the combinations of coefficients are

$$g_G(\psi,\varphi) = \left\{ W_{FLL_G}, W_{FHL_G}, W_{FLH_G}, W_{FHH_G}, \right\} \tag{8.9}$$

$$g_B(\psi,\varphi) = \left\{ W_{FLL_B}, W_{FHL_B}, W_{FLH_B}, W_{FHH_B}, \right\} \tag{8.10}$$

Equations (8.8)–(8.10) give the individual fused components of a color image.

8.3 Security of fused data/images

Color is said to be an effective descriptor. In a color image lots of extra information is contained, which can be used to simplify image analysis. To describe any particular color, three independent quantities are used. They are *hue*, *saturation*, and *brightness*. The *hue* is determined by the dominant wavelength. The *saturation* is determined by the excitation purity. It depends on the amount of white light mixed with the *hue*. The *brightness*, or *luminance*, is determined by the perception of the color, and is therefore psychological. It is assumed that different hues and luminance of colors might convey significant information therefore a color image plays an important role in our society.

Since a color image provides more information than a gray scale or binary image, the security of the color image has drawn the attention of the research community. The additional color information could contribute to the higher level of security. A review on the optical security techniques report that most of the literature deals with either binary or gray-scale images. The color image encryption-decryption techniques have been reportedly based on the use of a monochromatic light source, where original color information is not preserved. Therefore, the security of high-resolution color images is a challenging task and is of growing interest.

A color image encryption in the DRPE framework appeared in 1999 [5], in which the existing encryption method for gray-scale images was applied. A color image was converted to its indexed image format before encryption. During decryption, the color image was recovered after converting the decrypted indexed image back to its RGB format. Since color and shape information were combined together, it was claimed that better verification performance could be achieved. In DH, mostly a single wavelength is used for capturing holograms but, if multiple wavelengths are used then a color object can be reconstructed. This approach was applied for recording two holograms using red and green wavelengths and then they were fused to reconstruct a color image [6]. WT has good local optimization and multi-resolution analysis features, which attract its suitability in optical information processing applications. Therefore, using discrete WT the idea has been extended to multi-spectral holographic 3D image fusion [7]. In this study, a comparison of fusion results among Gaussian, Laplacian pyramid, and discrete WT fusion methods were carried out and it was claimed that discrete WT has an advantage over other methods. Discrete WT offers flexibility in controlling low as well as high frequency components, which help improve the fused image quality. Using a tunable solid-state pumped laser, 11 holograms were captured for wavelengths from 567 to 613 nm and fused together. Fused multi-wavelength hologram reconstructed images with multi-spectral information.

In case of color image encryption, it has been observed that when an encrypted image is decrypted with the wrong keys, there is a possibility of some traces of information. Therefore, to strengthen the level of security, encryption of color images using FRT was proposed, which used three channels corresponding to red, green, and blue color components [8]. A spectral image fusion analysis based on the discrete cosine transform has been reported, in which a specific spectral filtering method was used. Since compression helps to reduce the file size, a compression procedure based on adapted spectral quantization was suggested by encoding each frequency component with an optimized number of bits in the discrete cosine domain [9].

Most of the color image encryption schemes belong to the three-channels processing in which each channel is encrypted or decrypted independently. Such systems are considered as complex systems. To reduce the complexity in terms of processing, various single-channel color image encryption schemes have been proposed. A fusion technique for color images using WT for security and hiding applications has been reported [10]. The different components of a color image were fused together using discrete WT. The fused components were encrypted using the Fresnel domain amplitude- and phase-truncation approach. For the purpose of disguising information of the input image to an attacker, the encrypted components were transformed into different cover images. Further, 3D scene acquisition via reconstruction with multispectral information using computational integral imaging has been reported [11]. In this method, sensors covered with a Bayer color filter array captured elemental images at different visible spectral bands and subsequently, DRPE technique was applied.

Further, a color image-based authentication scheme was reported using multi-spectral photon counting imaging and DRPE [12]. An input multispectral color image was down-sampled into a Bayer-pattern image. Red, green, and blue color samples in the Bayer image were encrypted and the amplitude part of the resulting image was photon counted. Multispectral 3D photon-counted integral imaging has been further combined with Hartley transform and optical interferometry for security application [13]. The method was demonstrated with spatially and tempo-rally incoherent illumination. A multispectral cryptosystem has been demonstrated that exploits only few random white noise samples for decryption [14]. In the scheme, one of the two down-sampled images was converted into the phase function after shuffling through Arnold transformation and the other image was modulated as an amplitude-based image. Further, DRPE in FRT domain was applied for encryption. The encrypted complex data was randomly sampled via compressive sensing and with only 25% of the sparse white noise samples decryption was achieved.

Multispectral images have many applications including medicine and remote sensing for analysis. They are also used for reproducing the color of RGB images under different illumination. Highly accurate color reproduction can find applica-tions in many areas including visual communication systems for telemedicine and home care, online shopping, and digital archives. The development of a multi-spectral camera and display are important for practical utility. Such a camera has been reported, which converts a multispectral image to RGB and the multispectral display includes the inverse conversion process [15].

Since keys in the DRPE system have a noise-like distribution, which may catch people's attention, transmission through communication channels could bring more attacks. This issue was addressed through support constraints in an iterative encryption scheme. The axial translation of the CCD camera using amplitude- and phase-truncation provided support constraints. Also, a modified fusion technique was applied in the WT domain [16]. In a further study, the fusion technique for generating asymmetric keys for securing multiple images has been reported [17]. An input image was analytically encoded into two phase-only masks (POMs) by modulating the input image with an RPM. The process was repeated for three different images and one of the POMs corresponding to each input image was stored as a phase key and the other set of phase masks were fused together, which gave the encrypted image.

An optical hyperspectral image cryptosystem using improved Chirikov mapping in the gyrator domain has been reported [18]. This system can hide spatial and spectral information simultaneously. A hyperspectral image is converted into binary formats and then extended into a 1D array. Using Chirikov mapping, a position array is generated. Recently, reconfigurable MEMS Fano metasurfaces for performing logic operations at terahertz frequencies have been reported. It has been proposed that the XOR metamaterial gate may possess a potential application in cryptographically secured terahertz wireless communication networks [19].

8.4 Asymmetric cryptosystems with fused color components

In this section, a color image encryption under the framework of asymmetric cryptosystem is discussed. The approach is amplitude- and phase-truncation of the complex spectrum. First a color image is split into red, green, and blue color components. The fusion process based on discrete WT as discussed in section 8.2 has been followed. The algorithm for encryption of one of the color components, red, has been depicted in figure 8.3. The fused red component, $g_R(\psi,\Phi)$, is modulated with an RPM, $R_R(\psi,\Phi)$ and the resultant is Fresnel propagated for distance d_R. The obtained spectrum is phase-truncated (PT) and amplitude-truncated (AT).

$$E_{R_A}(u,\upsilon) = \text{PT}\left\{\mathfrak{J}_\lambda^{d_R}[g_R(\psi,\psi) \times R_R(\psi,\psi)]\right\} \qquad (8.11)$$

$$k_{R_P}(u,\upsilon) = \text{AT}\left\{\mathfrak{J}_\lambda^{d_R}[g_R(\psi,\varphi) \times R_R(\psi,\varphi)]\right\} \qquad (8.12)$$

For the retrieval of the red color component, the decryption key, $k_{R_P}(u,\upsilon)$, is generated. Similarly, the fused green color component is encrypted with Fresnel propagated distance d_G and using different RPM. Thus, correspondingly PT and AT operated values are obtained.

$$E_{G_A}(u,\upsilon) = \text{PT}\left\{\mathfrak{J}_\lambda^{d_G}[g_G(\psi,\varphi) \times R_G(\psi,\varphi)]\right\} \qquad (8.13)$$

$$k_{G_P}(u,\upsilon) = \text{AT}\left\{\mathfrak{J}_\lambda^{d_G}[g_G(\psi,\varphi) \times R_G(\psi,\varphi)]\right\} \qquad (8.14)$$

For the retrieval of the green color component, the decryption key, $k_{G_P}(u,\upsilon)$, is generated. Similarly, the fused blue color component is encrypted with Fresnel propagated distance d_B and using different RPM. Thus, correspondingly PT and AT operated values are obtained.

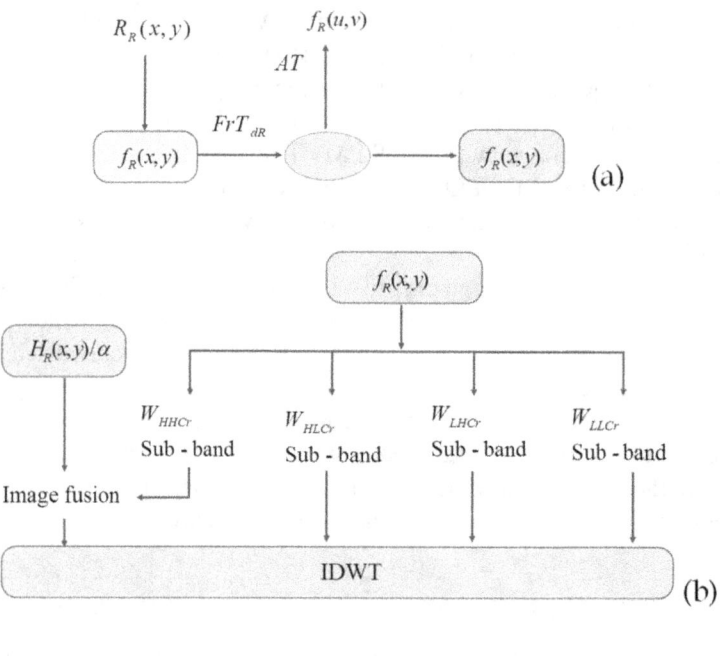

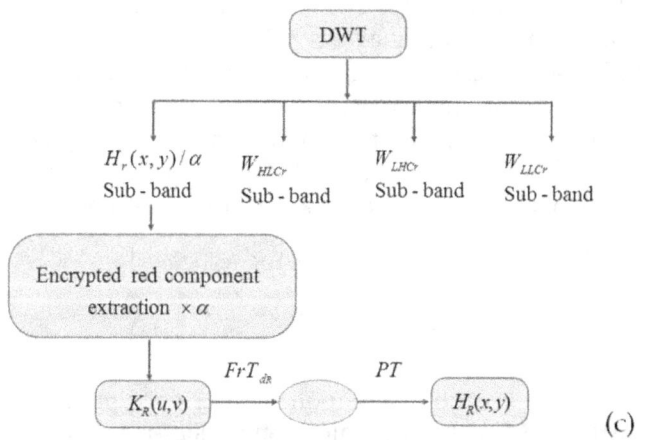

Figure 8.3. (a) Block diagram for asymmetric cryptosystem, (b) DWT-based image encryption, and (c) the decryption process.

$$E_{B_A}(u,v) = \text{PT}\left\{ \mathfrak{I}_\lambda^{d_B}[g_B(\psi,\varphi) \times R_B(\psi,\varphi)]\right\} \qquad (8.15)$$

$$k_{B_P}(u,v) = \text{AT}\left\{ \mathfrak{I}_\lambda^{d_B}[g_B(\psi,\varphi) \times R_B(\psi,\varphi)]\right\} \qquad (8.16)$$

For the successful retrieval of the blue color component, the decryption key, $k_{B_P}(u,v)$, is generated. The simulation results from [10] have been reproduced with permission. Figure 8.4(a) shows the input color image of size $256 \times 256 \times 3$ pixels.

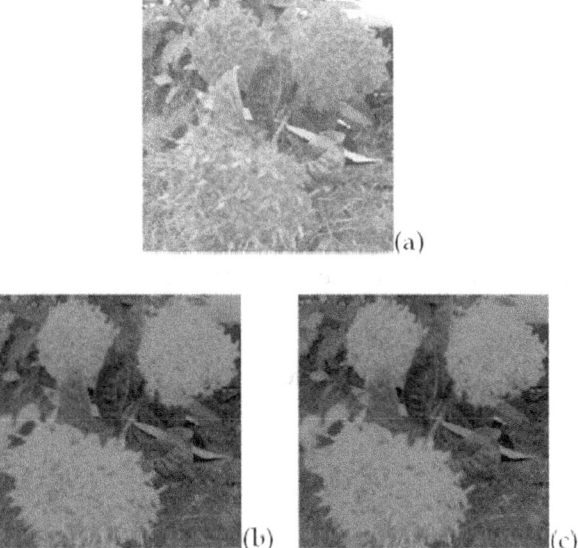

Figure 8.4. (a) Input color image of size 256 × 256 × 3 pixels, (b) the red component corresponding to $\lambda_R = 632.8$ nm, and (c) the red component corresponding to $\lambda_R + \delta\lambda_R = 631.8$ nm. Reprinted with permission from [10]. © The Optical Society.

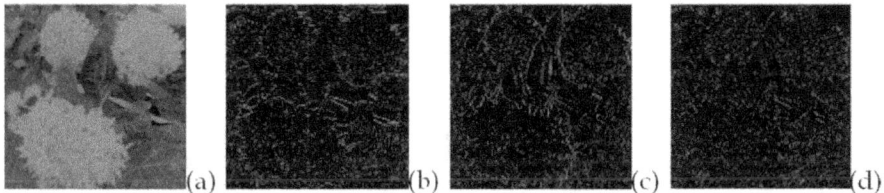

Figure 8.5. (a)–(d) Four coefficients of discrete WT of the reconstructed red component corresponding to wavelength, λ_R. Reprinted with permission from [10]. © The Optical Society.

Figures 8.4(b) and (c) show the red components corresponding to $\lambda_R = 632.8$ nm and $\lambda_R + \delta\lambda_R = 631.8$ nm, respectively. Similarly, green and blue color components corresponding to $\lambda_G = 532$ nm and $\lambda_G + \delta\lambda_G = 531$ nm, and $\lambda_B = 475$ nm and $\lambda_B + \delta\lambda_B = 476$ nm, respectively are generated. For image fusion, discrete WT has been applied. Figures 8.5(a)–(d) show all the four coefficients of discrete WT of the reconstructed red component corresponding to wavelength λ_R. Similarly, four coefficients of discrete WT of reconstructed red component corresponding to wavelength $\lambda_R + \delta\lambda_R$ were generated and the components were fused together. After achieving the fusion of wavelet coefficients, all color components in WT domain undergo an amplitude- and phase-truncation-based encryption process for generating the asymmetric (decryption) key and encrypted image for the red component.

To further enhance the level of security and employ more images/data simultaneously of encryption, a fusion technique has been proposed. The method helps to generate asymmetric keys [17]. A Laplacian pyramid fusion technique is a method in which low and high frequency components are merged together. The technique has been used in securing multiple images/data. Each input image is analytically encoded into two POMs; $M_1(\xi,\eta)$ and $M_2(\xi,\eta)$ in FRT (of order $\alpha_{n=1,2,3,...}$) domain. One of the POMs is stored as a security key while the other one is discrete wavelet transformed. The POMs corresponding to each input image are fused together using single level discrete WT for obtaining the encrypted image. The position multiplexing of wavelet coefficients, $[W_{FLL}, W_{FHL}, W_{FLH}, W_{FHH}]$ gives the encrypted image. The algorithms for asymmetric encryption-decryption have been shown in figures 8.6(a) and (b). For decryption, through position demultiplexing wavelet coefficients are separated. The wavelet coefficients corresponding to the ith image, behaving as asymmetric keys are generated.

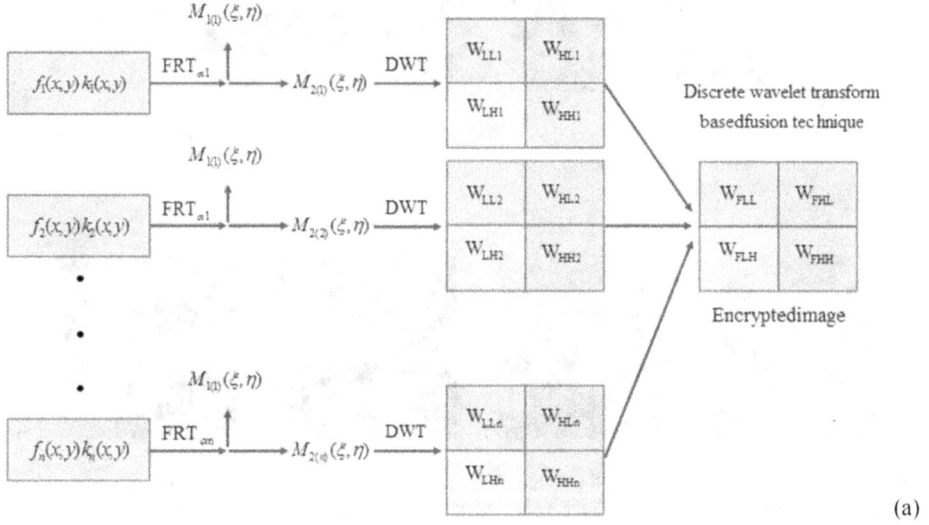

(a)

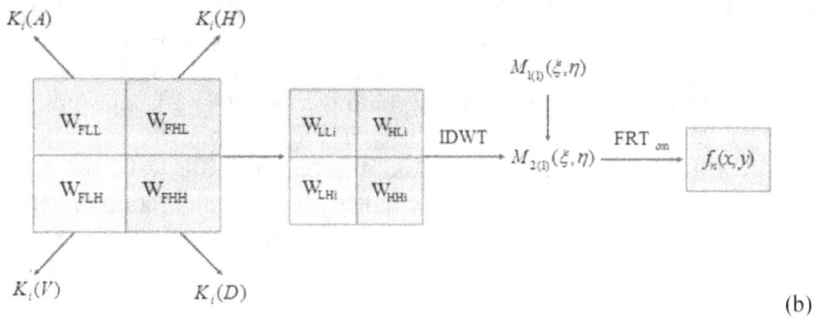

(b)

Figure 8.6. (a) Block diagram for the multiple image encryption scheme using the Laplacian pyramid fusion technique through discrete wavelet transform and (b) the block diagram for decryption.

$$k_{i(A)} = \frac{W_{LLi}}{AVG(W_{FLL1}, W_{FLL2},\ldots\ldots\ldots W_{FLLi}\ldots\ldots,W_{FLLn})} \qquad (8.17)$$

$$k_{i(H)} = \frac{W_{HLi}}{AVG(W_{FHL1}, W_{FHL2},\ldots\ldots\ldots W_{FHLi}\ldots\ldots,W_{FHLn})} \qquad (8.18)$$

$$k_{i(V)} = \frac{W_{LHi}}{AVG(W_{FLH1}, W_{FLH2},\ldots\ldots\ldots W_{FLHi}\ldots\ldots,W_{FLHn})} \qquad (8.19)$$

$$k_{i(D)} = \frac{W_{HHi}}{AVG(W_{FHH1}, W_{FHH2},\ldots\ldots\ldots W_{FHHi}\ldots\ldots,W_{FHHn})} \qquad (8.20)$$

where $k_{i(A)}$, $k_{i(H)}$, $k_{i(V)}$ and $k_{i(D)}$ represent the approximation, horizontal, vertical and diagonal keys corresponding to the ith image.

8.5 Color image encryption using XOR operation with LED

The XOR operation is a popular technique in optical cryptography. Through 2D XOR operation each pixel of a binary plaintext is individually encrypted using a key with sufficient randomness. XOR-based encryption of a QR code can be performed using an incoherent light source such as a light emitting diode (LED) through polarization encoding [20]. Use of an LED source improved the quality of decryption. Linear and inverse properties of XOR operation provide a simpler way of encryption but leads to security leakage. In an attempt to improve its effectiveness and strength, XOR encryption has been carried out using QR codes and the phase value distributions of plaintext evaluated using a phase retrieval algorithm [21]. It introduces randomness in the system and prevents tracing of the plaintext with partial information of the keys. For a color image, the bit-planes of each color component are individually encrypted and processed to obtain the key and the ciphertext.

Figure 8.7 presents the simulation results of a XOR encryption process for a color image. The plaintext shown in figure 8.7(a) has been decomposed into its color components, which are converted into phase value distributions related by the Fourier transformation. Figures 8.7(b) and (c) show the green component and corresponding phase value distribution. Each bit-plane of phase value distributions are encrypted using QR codes. As an example, XOR operation of the 8th bit-plane with a random QR code to obtain the corresponding bit-plane of the key has been shown in figures 8.7(d)–(f), respectively. Further, the QR codes and resulting images are processed to obtain the ciphertext as shown in figure 8.7(g). The XOR-operated images converted into a key, and successfully decrypted image obtained after using correct key have been shown in figures 8.7(h) and (i), respectively.

XOR operation based on polarization encoding has been optically implemented using a pair of twisted nematic liquid crystal SLMs placed between crossed polarizers through an incoherent light illumination, as shown in figure 8.8. The SLMs were biased at two distinct voltages, denoted by low and high voltages. At low

Figure 8.7. Simulation results. (a) Plaintext, (b) the green component of the plaintext, (c) phase value distribution for the green component, (d) one of the bit-planes of phase value distribution, (e) one of the QR codes, (f) the XOR-operated image, (g) QR codes converted into ciphertext, (h) XOR-operated images converted into a key, and (i) decrypted image obtained after using the correct key.

voltage, the SLM rotates the light polarization by 90° while at high voltage it simply passes the light. The analyzer decodes the polarization state of the output beam into intensity variations which gets recorded in a camera. In the encryption procedure, keys are generated by encoding the individual bit-planes of plaintext and QR codes onto the SLMs.

Experimental results for optical XOR operation for key generation performed using a white LED light source has been shown in figures 8.9(a)–(d). The intensity image of one of the bit-planes of phase value distribution of the green component has been shown in figure 8.9(a) and the similarly obtained image of one of the used QR codes has been shown in figure 8.9(b). The XOR-operated images of (a) and (b) for key generation (one of the bit-planes) have been presented in figure 8.9(c) and the processed image of (c) is shown in figure 8.9(d). For decryption, correct bit-planes of key and ciphertext are individually encoded onto the SLMs to obtain the bit-planes of plaintext.

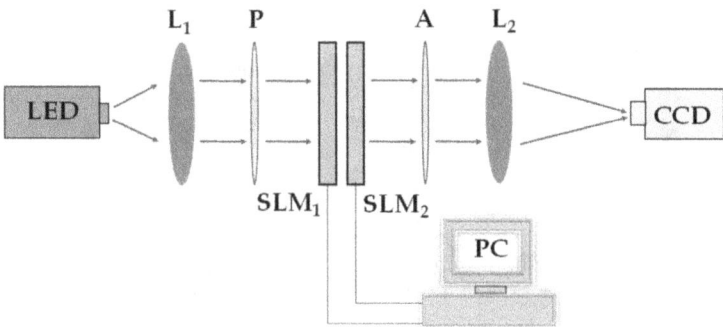

Figure 8.8. Schematic of the experimental setup. LED: light emitting diode, L: lens, SLM: spatial light modulator, P: polarizer, A: analyzer, CCD: charge-coupled device camera, PC: personal computer.

Figure 8.9. Experimental results for optical XOR operation for key generation performed using a white LED light source. (a) Intensity image of one of the bit-planes of phase value distribution of the green component, (b) similarly, the image of one of the used QR codes, (c) the XOR-operated image of (a) and (b) for key generation (one of the bit-planes), and (d) the processed image of (c).

The experimental results for optical XOR operation for decryption have been shown in figures 8.10(a)–(d). Figure 8.10(a) presents the XOR-operated image obtained after using the correct keys for one of the bit-planes of the plaintext with white LED and figure 8.10(b) presents the processed image of (a). Figure 8.10(c) shows the XOR-operated image obtained after using correct keys for one of the bit-planes of the plaintext with green LED, and figure 8.10(d) presents the XOR-operated image obtained after using the correct keys for one of the bit-planes of the

Figure 8.10. Experimental results for optical XOR operation for decryption. (a) XOR-operated image obtained after using correct keys for one of the bit-planes of the plaintext with white LED, (b) the processed image of (a), (c) the XOR-operated image obtained after using the correct keys for one of the bit-planes of the plaintext with green LED, and (d) XOR-operated image obtained after using the correct keys for one of the bit-planes of the plaintext with laser ($\lambda = 532$ nm).

plaintext with laser ($\lambda = 532$ nm). It can be seen that with the use of a laser source, the output contains speckles while with LED the output is much better with less noise. Therefore, the use of LED and a QR code not only reduces the damage caused by speckles but also overcomes the noise problem.

Similarly, other such images can be obtained for each component of plaintext which could be combined to decrypt the color image. Since applications of image fusion are diverse and growing, the security of multispectral data is also becoming increasingly important. Wavelet transform has peculiar properties and is considered as a potential candidate for not only achieving fusion but security also. Multispectral images/data can also be secured using a QR code and with partial coherent light sources such as LEDs.

References

[1] Sadjadi F 2002 Invariant algebra and the fusion of multi-spectral information *Inf. Fusion* **3** 39–50

[2] Blum R S and Liu Z (ed) 2005 *Multi-Sensor Image Fusion and Its Applications* (Boca Raton, FL: CRC Press)

[3] Zyczkowski M, Szustakowski M, Ciurapinski W, Kastek M, Dulski R, Karol M, Kowalski M and Markowski P 2013 Multispectral solutions in surveillance systems: the need for data fusion *WIT Trans. Built Environ.* **134** 285–93

[4] Young R K 1993 *Wavelet Theory and Its Applications* (Amsterdam: Kluwer)

[5] Zhang S and Karim M A 1999 Color image encryption using double random phase encoding *Microw. Opt. Technol. Lett.* **21** 318–23

[6] Javidi B, Ferraro P, Hong S H, DeNicola S, Ginizio A, Alfieri D and Pierattini G 2005 Three-dimensional image fusion by use of multiwavelength digital holography *Opt. Lett.* **30** 144–6

[7] Javidi B, Do C M, Hong S H and Nomura T 2006 Multispectral holographic three-dimensional image fusion using discrete wavelet transform *J. Disp. Technol.* **2** 411–7

[8] Joshi M, Chandrashakher and Singh K 2007 Color image encryption and decryption using fractional Fourier transform *Opt. Commun.* **279** 35–42

[9] Alfalou A, Brosseau C, Abdallah N and Jridi M 2011 Simultaneous fusion, compression, and encryption of multiple images *Opt. Express* **19** 24023–9

[10] Mehra I and Nishchal N K 2014 Image fusion using wavelet transform and its application to asymmetric cryptosystem and hiding *Opt. Express* **22** 5474–82

[11] Muniraj I, Kim B and Lee B G 2014 Encryption and volumetric 3D object reconstruction using multispectral computational integral imaging *Appl. Opt.* **53** G25–32

[12] Yi F, Moon I and Lee Y H 2014 A multispectral photon-counting double random phase encoding scheme for image authentication *Sensors* **14** 8877–94

[13] Muniraj I, Guo C, Lee B-G and Sheridan J T 2015 Interferometry based multispectral photon-limited 2D and 3D integral image encryption employing the Hartley transform *Opt. Express* **23** 15907–20

[14] Rawat N, Kim B, Muniraj I, Situ G and Lee B G 2015 Compressive sensing based robust multispectral double-image encryption *Appl. Opt.* **54** 1782–93

[15] Shinoda K, Watanabe A, Hasegawa M and Kato S 2015 Multispectral information hiding in RGB image using bit-plane-based watermarking and its application *Opt. Rev.* **22** 469–76

[16] Mehra I and Nishchal N K 2015 Optical asymmetric watermarking using modified wavelet fusion and diffractive imaging *Opt. Lasers Eng.* **68** 74–82

[17] Mehra I and Nishchal N K 2015 Wavelet-based image fusion for securing multiple images through asymmetric keys *Opt. Commun.* **335** 153–60

[18] Chen H, Tanougast C, Liu Z, Blondel W and Hao B 2018 Optical hyperspectral image encryption based on improved Chirikov mapping and gyrator transform *Opt. Lasers Eng.* **107** 62–70

[19] Manjappa M, Pitchappa P, Singh N, Wang N, Zheludev N I, Lee C and Singh R 2018 Reconfigurable MEMS Fano metasurfaces with multiple-input-output states for logic operations at terahertz frequencies *Nature Commun.* **9** 1–10

[20] Kumar P and Nishchal N K 2019 Enhanced exclusive-OR and quick response code-based image encryption through incoherent illumination *Appl. Opt.* **58** 1408–12

[21] Kumar P, Fatima A and Nishchal N K 2019 Image encryption using phase-encoded exclusive-OR operations with incoherent illumination *J. Opt.* **21** 065701

IOP Publishing

Optical Cryptosystems

Naveen K Nishchal

Chapter 9

Chaos-based information security

9.1 Introduction

Randomness is observed in many physiological systems that are found abundantly in nature. A complex system is said to be random if it possesses a large number of degrees of freedom which may not be directly observed but may be revealed by observing the fluctuations. Of late, researchers from various fields have shown that the spectrum observed in a dynamical system may not be, in a lot of cases, distinguished by a spectrum of purely random activity. This phenomenon is called chaotic behavior and is associated with the science of nonlinear dynamics. The complex and unpredictable behaviour of a chaotic phenomenon is key to non-linearity. A nonlinear system does not obey the law of superposition as linear systems do. It is important to state that chaotic behaviour is not always present in every nonlinear system. This implies that nonlinearity is a necessary condition for a chaotic system but it is not a sufficient condition. The chaotic behaviour of a nonlinear dynamical system in principle is the complex trajectory traced which follows non-periodic behavior that mimics randomness. A nonlinear system is generally described by a set of differential equations. To be called chaotic, the dimension of the differential equation should be at least three and should consist of nonlinear terms. Another fact worth mentioning is that chaotic behaviour is only expected in a short range of control parameters defined for the system [1–4].

The sensitivity to a set of initial conditions makes a chaotic system unique. The trajectories traced by a slight change in the initial conditions are entirely different and the measured deviation increases exponentially with the time evolution. The overall chaoticity is measured by a metric known as the *Lyapunov exponent*. The number of the Lyapunov exponent is the same as the dimension of a dynamical system. For a nonlinear system to be chaotic at least one of the Lyapunov exponents should be positive. A system possessing chaotic nature may be characterized by the following properties:

doi:10.1088/978-0-7503-2220-1ch9

- **Extreme sensitiveness to initial conditions**: a nonlinear dynamical system is expressed by a set of differential equations, which is deterministic if the previous state variables are fully known. Otherwise, it is impossible to predict the future states. For the same set of control parameters, a slight change in the initial values results in totally different trajectories, which exponentially deviate with time.
- **Ergodicity**: ergodicity relates to a process in which every sequence equally represents the whole. A chaotic system is said to be ergodic if the trajectory traced in phase space is in close proximity to the earlier states. It also indicates that the system traverses after repeated time iterations to a constricted set of points and produces an attractor. The density of these set of points are invariant with time and have potential applications in cryptography.
- **Mixing**: the time evolution of the trajectory being traced is spread over the full phase space for a small range of initial conditions. Irrespective of the interval selected for the initial conditions, the spreading is confined to that part of phase space where there is a confinement of trajectory in an asymptotic manner. Thus, there exists an overlap of area in the phase space of the attractor spatially.

9.2 Chaos and cryptography

The strength of a cryptographic algorithm lies in the selection of the keys. The keys are secret parameters and should not be known to an adversary. A chaotic system is extremely sensitive to initial conditions. This property has been widely explored in developing cryptographic algorithms [5–8]. A chaotic signal is intrinsically dedicated to encryption. The dynamics of the chaotic signal is similar to that of a white noise and they are deterministic. The complicated dynamics of a chaotic system can be governed by a nonlinear differential equation, which is used during the encoding-decoding process. The conventional data encryption techniques, which use an encryption key to control the cryptographic algorithm, are slow. On the other hand, chaos is used as a coding key embodied directly in the structure of the carrier and is fast. For encryption, the chaos-based approaches are chaotic in intensity and chaotic in wavelength. Although it is natural to look for nonlinearities induced by optical power when discussing chaos in optics. In optical cryptosystems, wavelength is preferred than power to be used for producing optical chaos. Chaos in wavelength has the advantages of high accuracy and high reliability of chaos control. Generating chaos in wavelength relies on the wavelength agility of a laser diode with a feedback loop and a nonlinear element in wavelength [5, 6].

Chaos-based encryption methods are considered to have a good combination of speed, high security, complexity, reasonable computational cost, and computational power. Such a system electrically superimposes the information signal to the chaotic signal that propagates into the feedback loop. Chaos-based image encryption algorithms may be implemented in both digital and optical domains [4–8].

The chaotic attractor is sensitive to the values set as initial conditions. The values generally used are decimals, which may be non-terminating, non-repeating or any other combination. The sensitivity can be ascertained to the fact that even a slight change in the digits after the decimal place results in entirely different trajectories. It is therefore evident that a large key space would make the cryptographic algorithm stronger. The initial values themselves may serve as the keys for encryption and decryption making the cryptography symmetric.

A general image encryption scheme has two important stages: confusion and diffusion. The confusion stage randomly changes the location of pixels without changing its value, while the diffusion stage alters the values of the pixels. For an encryption algorithm to perform well, a single digit of the key should be able to influence a large number of digits of the plaintext. Not all dynamical systems are candidates for cryptographic applications but only those can be used that are able to retain their qualitative properties even in small perturbations. In other words, chaotic systems with stable attractors with good diffusion effect can only be employed in cryptographic algorithms. It is also stated that chaos may be a necessary condition but in totality it is not a sufficient condition for a good encryption algorithm. A good encryption algorithm should have the properties of confusion, diffusion, and sensitivity to changes in key parameters and the original plaintext. To resist cryptanalysis attacks, a chaotic system should be used in the simplest and irreducible form.

For chaos-based encryption, it has been reported that time-delayed nonlinear systems can produce chaos of an extremely high dimension and such systems are capable of a high degree of security [7]. An image encryption scheme employing a chaotic logistic map, an 80-bit secret key provided externally and two logistic maps has been proposed in the digital domain. The scheme used eight different types of operations in the encryption stage [8]. An image encryption method using FRT and chaos has also been reported, in which chaotic RPMs were generated using iterative chaos functions [9]. In this scheme, three types of chaos functions were used: the logistic map, the tent map, and the Kaplan–Yorke map.

For the purpose of secure image transmission, an image encryption scheme based on the juxtaposition of sections of the image in multi-parameter discrete FRT domain has been reported. In this method, the alignment of sections was determined through chaotic logistic maps. This chaos-based pixel scrambling method did not require any phase keys [10]. Further, a color image encryption using chaotic scrambling and FRT has been reported. The hue, saturation and intensity components of a color image were converted into amplitude information or phase information to implement a single-channel encryption. The chaotic scrambling caused disturbance in both the spatial and frequency domain [11]. Since the diffusion process enhances the resistance to statistical and differential attack, an image encryption method based on a permutation-diffusion structure has been reported, in which for permutation and diffusion processes chaotic maps were employed [12]. The permutation process used a generalized Arnold map to generate a chaotic map. To improve the diffusion effect, one generalized Arnold map and one generalized Bernoulli shift map were used for generating two pseudo-random gray value sequences. These two sequences were used to modify the pixel gray values sequentially. Further, an encryption algorithm based on iterating chaotic maps has been reported, in which pseudo-random sequences were generated by a group of 1D chaotic maps [13]. The method was applied securing gray-scale and color images.

In permutation-diffusion type systems, the diffusion process is very time consuming. Actually in real number arithmetic operation and subsequent quantization, considerable amount of computation load is required for key stream generation. To accelerate the spreading process, a chaos-based image encryption using bidirectional diffusion strategy has been reported [14]. In this method, a plaintext dependent chaotic orbit turbulence mechanism has been used in the diffusion process through the perturbation of control parameter of the chaotic system according to the cipher-pixel. An algorithm for color image encryption using chaos-based cyclic shift and multiple order discrete fractional cosine transform has been reported [15]. In this case, the fractional orders were determined by a Chirikov standard chaotic map. Further, a CGH and chaos-based encryption-decryption technique has been reported, in which random phase arrays were constructed by a chaotic sequence of the deterministic nonlinear system [16]. The basic advantage of the technique as reported was that the initial value of the chaotic function was enough to generate the random phase arrays for decryption. An image encryption technique based on the chaotic Baker map and DRPE has also been reported, which was implemented in two layers to enhance the level of security. The first layer was a pre-processing layer, which was performed with the chaotic Baker map on the original image and in the second layer classical DRPE was utilized [17]. Further, an encryption algorithm based on a hyper-chaotic system has been reported, in which for the initial conditions of the hyper-chaotic system 256-bit long external secret key was used [18].

Combining compressive sensing (CS) and chaos, a method for simultaneously achieving compression, fusion, and encryption of multi-modal images has been reported [19]. The method used a logistic map for a key-controlled pseudo-random measurement matrix, which was fused through the adaptive weighted fusion rule. The fused measurement was encrypted using the pixel exchange technique and iterative procedure in the FRT domain. The potential applications of the proposed method as envisaged were in telemedicine, military surveillance, etc. In such applications, multi-modal medical images need to be fused to integrate complementary information and to be secured in order to protect the privacy and to reduce the amount of transferred data compression is very important.

Chaos has been combined with the phase retrieval algorithm in the Fresnel domain to achieve multiple binary image encryption [20]. In this study, each input image was encrypted into a Fresnel diffraction intensity pattern using two chaos-generated RPMs and the intensity patterns were partially selected by different binary masks. Such patterns were then combined together to get a single intensity pattern, which was called an encrypted image. For decryption, the phase retrieval algorithm was applied with support constraint in the output plane and along with median filtering. Further, a study on acousto-optic chaos for its application in secure free-space communication has been reported [21]. In this case, radio frequency chaos from a hybrid acousto-optic feedback device was used for encryption. Recently, a QR code and optical interference-based asymmetric encryption scheme have been reported, in which the input image was scrambled using an Arnold cat map [22].

9.3 Chaos functions

In this section, some of the chaos functions are discussed, which have been used in chaos-based cryptosystems. It has been demonstrated that securing information via chaos waves leads to high robust transmission and recovery. Implementation of digital chaos implies finite numbers either with floating-point arithmetic, fixed point arithmetic or other arithmetic at a reasonable digital size. It should be noted that chaotic maps are discrete by nature.

The **Logistic map** is a 1D chaos function. It is defined as [9]

$$f(x) = l \, x \, (1 - x) \tag{9.1}$$

This function is bounded for $0 < l < 4$ and can be written in the iterative form as

$$x_{n+1} = l \, x_n \, (1 - x_n) \tag{9.2}$$

Here, x_0 gives the initial value, which is also known as the seed value for the chaos function.

The **Tent map** is a 1D chaos function. It is defined as [9]

$$f(x) = t \, x \qquad \text{for } 0 \leqslant x \leqslant 0.5 \tag{9.3}$$

$$f(x) = t(1 - x) \qquad \text{for } 0.5 \leqslant x \leqslant 1 \tag{9.4}$$

This function is bounded for $0 < t \leqslant 2$ and can be written in the iterative form as

$$x_{n+1} = t \, x_n \qquad \text{for } 0 \leqslant x_0 \leqslant 0.5 \tag{9.5}$$

$$x_{n+1} = t(1 - x_n) \qquad \text{for } 0.5 \leqslant x_0 \leqslant 1 \tag{9.6}$$

Here, x_0 gives the initial value.

The **Kaplan–Yorke map** is a 2D chaos function. It is defined as [9]

$$f(x) = a \, x \bmod 1 \tag{9.7}$$

$$f(y) = by + \cos(4\pi x) \tag{9.8}$$

This function is bounded for $0 \leqslant a \leqslant 2$ and $0 \leqslant b \leqslant 2$ and can be written in the iterative form as

$$x_{n+1} = a \, x_n \bmod 1 \tag{9.9}$$

$$y_{n+1} = by_n + \cos(4\pi x_n) \tag{9.10}$$

Here, x_0 gives the initial value.

The **Chirikov standard map** is a 2D chaotic map, which is used to shuffle the pixel positions of a plain image. This is an area-preserving chaotic map from a square with side 2π onto itself. It is defined as [14]

$$x_{i+1} = (x_i + y_i)\bmod 2\pi \tag{9.11}$$

$$y_{i+1} = \{y_i + c\sin(x_i + y_i)\}\bmod 2\pi \tag{9.12}$$

where c is the control parameter satisfying $c > 0$, and the ith states x_i and y_i both take real values in $[0, 2\pi)$ for all i. For $c = 0$, the Chirikov standard map is linear and periodic and quasi-periodic orbits also exist.

The **Chaotic Baker map** is a permutation-based tool, which performs randomization of a square matrix by changing the pixel positions. In a bijective manner it assigns a pixel to another pixel position [17]. The discrete chaotic Baker map is denoted by $B(b_1, b_2, \ldots b_k)$, where the sequence of k integers, $b_1, b_2, \ldots b_k$ is chosen such that each integer b_i divides M, and $M_i = b_1 + b_2 + \ldots + b_i$. The pixel at indices (l,s) with $M_i \leqslant l < M_i + b_i$ and $0 \leqslant s < M$ is mapped to

$$B_{(n_1,n_2,\ldots n_k)}(l,s) = \left[\frac{M}{b_i}(l - M_i) + s\bmod\frac{M}{b_i}, \frac{b_i}{M}\left(s - s\bmod\frac{M}{b_i}\right) + M_i\right] \tag{9.13}$$

The **spatiotemporal chaotic function**, which is based on a skew tent map is defined as [20]

$$x_{n+1}^i = (1 - s)f\left(x_n^i\right) + \frac{\varepsilon}{2}\left[f\left(x_n^{i-1}\right) + f\left(x_n^{i+1}\right)\right] \tag{9.14}$$

where x_n^i represents the state variable for the site i ($i = 1, 2, 3, \ldots N$) at time n ($n = 1, 2, 3, \ldots$). Here, N is the number of sites and s is a coupling constant. The function $f(.)$ is a skew tent map. It is defined as

$$f(x^i) = \begin{cases} \dfrac{x^{i-1}}{a}, & 0 \leqslant x^{i-1} < a \\[2mm] \dfrac{1 - x^{i-1}}{1 - a}, & a \leqslant x^{i-1} < 1 \end{cases} \tag{9.15}$$

The initial driven sequence x_0^i ($i = 1, 2, 3, \ldots N$) is generated by the Logistic map as defined in equation (9.1).

The **Arnold cat map** is a 2D chaotic map, which is invertible. When the Arnold cat map is applied to an input image, it randomizes the original organization of pixels. Its period is determined by the number of iterations. It is defined as [22]

$$\begin{bmatrix} x \\ y \end{bmatrix} = A\begin{bmatrix} x' \\ y' \end{bmatrix}\bmod(256) = \begin{bmatrix} 1 & a \\ b & ab + 1 \end{bmatrix}\begin{bmatrix} x' \\ y' \end{bmatrix}\bmod(256) \tag{9.16}$$

Here, a and b are positive integers and the operator mod denotes the modulus. This chaotic function is an area-preserving map since $\det(A) = 1$. (x',y') and (x,y) denotes the coordinates and transformed coordinates of pixels of an ($N \times N$) image, respectively.

9.4 Chaos-based optical asymmetric cryptosystem

It has been reported that encryption schemes based on the amplitude- and phase truncation approach are vulnerable to specific attack. Therefore, in order to add nonlinearity to the asymmetric schemes, chaotic systems have been unified. Of late, cryptosystems based on digital chaos have attracted more attention as compared to its analog version because of the advantages offered by discrete chaotic maps in terms of security, performance, flexibility, and cost. In this section, an optical asymmetric cryptosystem that combines the chaos with the amplitude- and phase-truncation approach and a QR code is discussed. For scrambling the pixel positions of an input image, the chaotic Arnold cat map has been applied, which is then divided into a number of pixel blocks. Each block is encoded into the corresponding QR code, which are then multiplexed to obtain a single binary matrix. Thus, the obtained matrix is further processed by an asymmetric cryptosystem in the Fresnel domain. The flowchart of the encryption process is shown in figure 9.1.

An input image denoted by function $f(x,y)$ is multiplied by an RPM and its Fresnel transformation is calculated.

$$F(u,v) = \frac{\exp(ikd)}{i\lambda d} \iint \{f(x,y) \times \exp[i2\pi R_1(x,y)]\}$$
$$\times \exp\left(\frac{ik}{2d}\right)[(u-x)^2 + (v-y)^2]\mathrm{d}x\mathrm{d}y \tag{9.17}$$

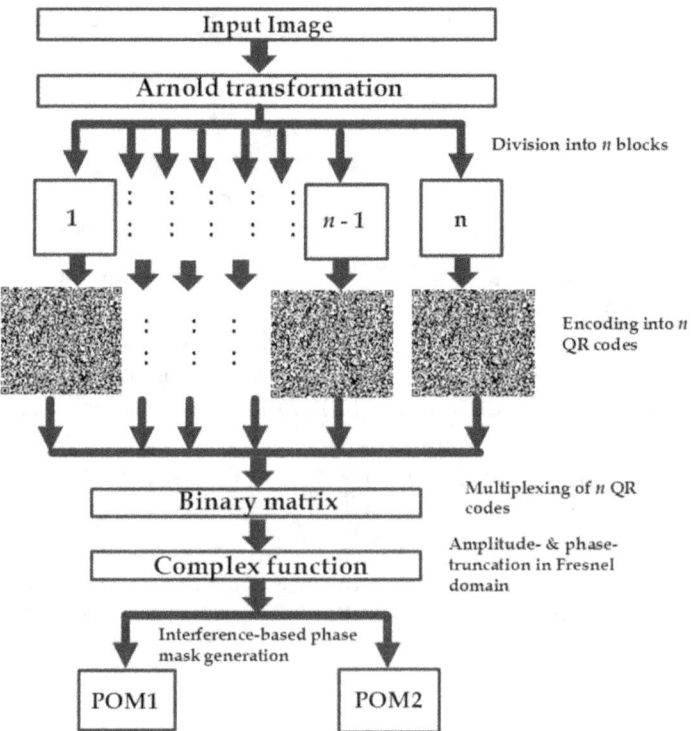

Figure 9.1. Flowchart for the image encryption process.

Here, (x,y) and (u,v) are the coordinates of the input and output planes, respectively. λ denotes the optical wavelength, $k = \frac{2\pi}{\lambda}$, and d denotes the free space propagation distance. The obtained spectrum $F(u,v)$ is phase-truncated (PT) and multiplied by another RPM and again Fresnel-transformed.

$$G(\xi,\eta) = \mathfrak{I}_\lambda^d[\text{PT}\{F(u,v)\} \times \exp[i2\pi R_2(u,v)]]] \tag{9.18}$$

Here, $\mathfrak{I}_\lambda^d$ denotes Fresnel transform operation. The obtained spectrum, $G(\xi,\eta)$, is amplitude- and phase-truncated.

$$G_P(\xi,\eta) = \text{AT}[G(\xi,\eta)] \tag{9.19}$$

$$G_A(\xi,\eta) = \text{PT}[G(\xi,\eta)] \tag{9.20}$$

The term 'AT' denotes the amplitude-truncation operation. The function, $G_P(\xi,\eta)$ serves as the decryption key. The 2nd decryption key is obtained as

$$K_D = \frac{AT[\mathfrak{I}_\lambda^d\{f(x,y) \times \exp[i2\pi R_1(x,y)]\}]}{PT[\mathfrak{I}_\lambda^d\{f(x,y) \times \exp[i2\pi R_1(x,y)]\}]} \tag{9.21}$$

The function, $G_A(\xi,\eta)$ is encoded into two analytically generated POMs by modulating this function with a new RPM.

$$H(\xi,\eta) = G_A(\xi,\eta)\exp[i2\pi\psi(\xi,\eta)] \tag{9.22}$$

The complex function, $H(\xi,\eta)$, is expressed as the interference of two POMs: POM_1 and POM_2 [22].

$$H(\xi,\eta) = \exp[i\text{POM}_1(x,y)] \times h(x,y,l) + \exp[i\text{POM}_2(x,y)] \times h(x,y,l) \tag{9.23}$$

where

$$h(x,y,l) = \frac{\exp\left(\frac{i2\pi l}{\lambda}\right)}{il\lambda} \exp\left[\frac{i\pi}{l\lambda}(x^2 + y^2)\right] \tag{9.24}$$

The function $h(x,y,l)$ represents the point pulse function of the Fresnel transform. The distance between phase mask and the output plane is given by l. Equation (9.22) can be expressed as

$$\exp[i\text{POM}_1(x,y)] \times h(x,y,l) + \exp[i\text{POM}_2(x,y)] \times h(x,y,l) = \mathfrak{I}^{-1}\left[\frac{\mathfrak{I}\{H(\xi,\eta)\}}{\mathfrak{I}\{h(x,y,l)\}}\right] \tag{9.25}$$

Introducing a new term,

$$D = \mathfrak{I}^{-1}\left[\frac{\mathfrak{I}\{H(\xi,\eta)\}}{\mathfrak{I}\{h(x,y,l)\}}\right] \tag{9.26}$$

Equation (9.24) can now be rewritten as

$$\exp[i\text{POM}_2(x,y)] = D - \exp[i\text{POM}_1(x,y)] \tag{9.27}$$

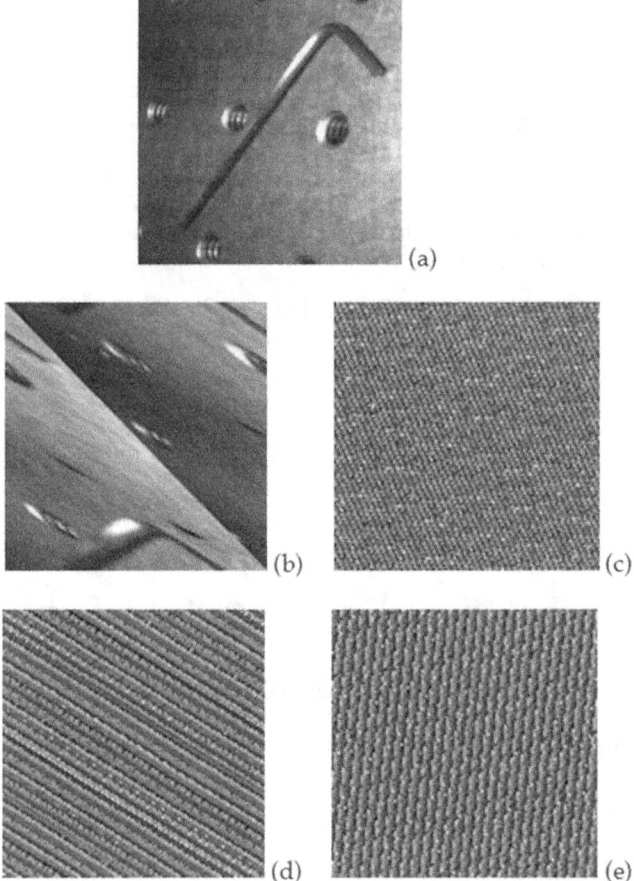

Figure 9.2. (a) Input image of an Allen key. The resultant image obtained after (b) one iteration of Arnold transform, (c) 10 iterations of Arnold transform, (d) 100 iterations of Arnold transform, and (f) 1000 iterations of Arnold transform.

Figure 9.3. POMs obtained as a result of encrypting the image of the Allen key: (a) POM, P_1, (b) POM, P_2, and (c) the decrypted Allen key obtained after using the correct decryption parameters.

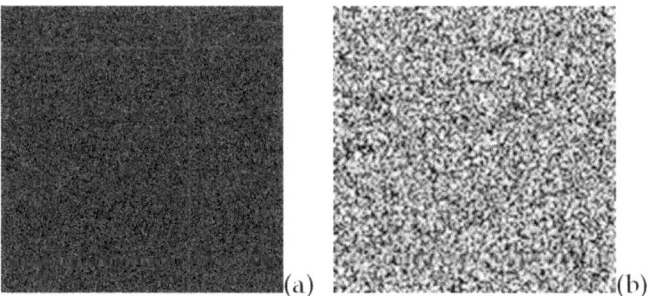

Figure 9.4. (a) Incorrect POM, P_1 and (b) the decrypted Allen key obtained after incorrect P_1 but correct P_2.

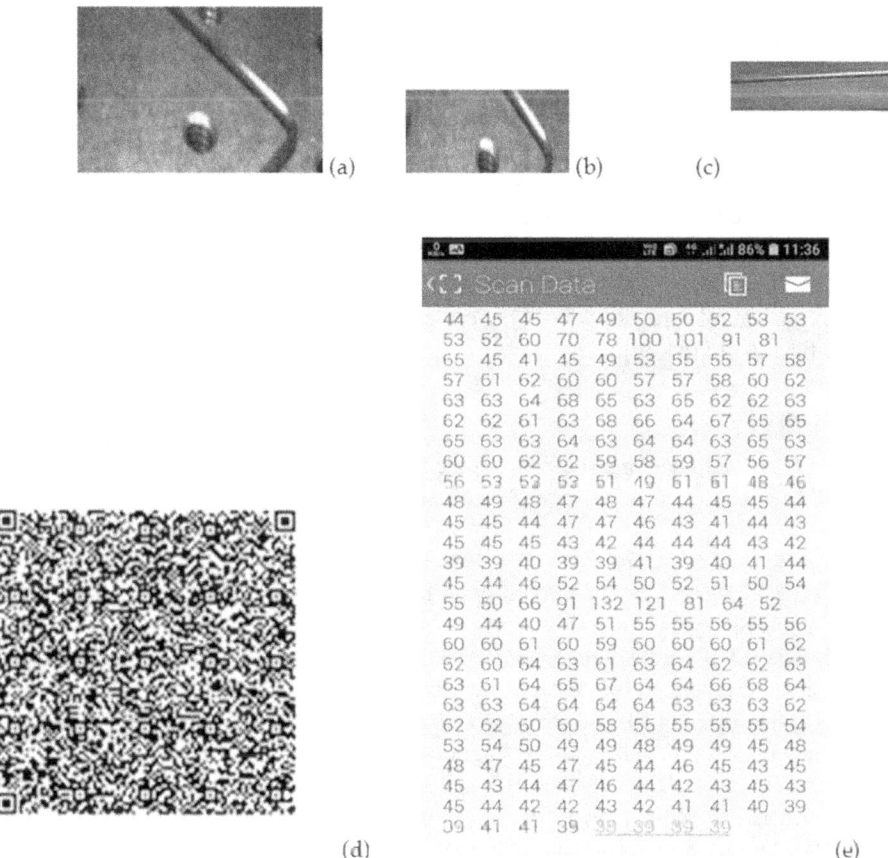

Figure 9.5. Partially decrypted image of the Allen key using (a) 32 QR codes, (b) 16 QR codes, (c) 8 QR codes, (d) the 25th QR code obtained during the decryption process, and (e) a screenshot of the decoded QR code data captured using a smartphone.

The phase-only information is present in the two POMs. This can be expressed as

$$|D - \exp\{i\mathrm{POM}_1(x,y)\}|^2 = [D - \exp\{i\mathrm{POM}_1(x,y)\}] \times [D - \exp\{i\mathrm{POM}_2(x,y)\}] = 1 \quad (9.28)$$

The POMs can now be expressed as

$$\mathrm{POM}_1(x,y) = \arg(D) - \arccos\left(\frac{\mathrm{abs}(D)}{2}\right) \quad\quad (9.29)$$

$$\mathrm{POM}_2(x,y) = \arg(D) - \exp\{i\mathrm{POM}_1(x,y)\} \quad\quad (9.30)$$

Here, $\arg(D)$ and $\mathrm{abs}(D)$ would give the phase and magnitude of the complex function, respectively.

In chaos-based cryptosystems, chaos is used as a coding key, which is directly embodied into the structure of the carrier and is considered highly robust and fast. Chaotic systems are inherently discrete in nature and suitable for secure transmission.

The simulation results based on the chaos-based asymmetric scheme are presented in figures 9.2–9.5. An input image is depicted in figure 9.2(a), and figures 9.2(b)–(d) show the images obtained after various iterations of Arnold transform. The analytically generated POMs are shown in figure 9.3(a) and (b), and the decrypted image obtained after using the correct decryption parameters is shown in figure 9.3(c). For successful decryption, both the POMs are required. If any of the POMs are incorrect, the obtained decrypted image remains a noisy pattern (figure 9.4). The decrypted images obtained with different numbers of QR codes and a screenshot of the decoded QR code data captured with a smartphone are shown in figures 9.5(a)–(e).

References

[1] Lorenz E N 1963 Deterministic nonperiodic flow *J. Atmos. Sci.* **20** 130–41
[2] Pecora L M and Carroll T L 1990 Synchronization in chaotic systems *Phys. Rev. Lett.* **64** 821–4
[3] Banerjee S, Yorke J A and Grebogi C 1998 Robust chaos *Phys. Rev. Lett.* **80** 3049–52
[4] Alligood K T, Sauer T D and Yorke J A 2001 *Chaos: An Introduction to Dynamical Systems* (New York: Springer)
[5] Goedgebuer J P, Larger L and Porte H 1998 Optical cryptosystem based on synchronization of hyperchaos generated by a delayed feedback tunable laser diode *Phys. Rev. Lett.* **80** 2249
[6] Larger L, Goedgebuer J P and Delorme F 1998 Optical encryption system using hyperchaos generated by an optoelectronic wavelength oscillator *Phys. Rev.* E **57** 6618–24
[7] Cuenot J B, Larger L, Goedgebuer J P and Rhodes W T 2001 Chaos shift keying with an optoelectronic encryption system using chaos in wavelength *IEEE J. Quant. Electron.* **37** 849–55
[8] Pareek N K, Patidar V and Sud K K 2006 Image encryption using chaotic logistic map *Image Vision Comput.* **24** 926–34
[9] Singh N and Sinha A 2008 Optical image encryption using fractional Fourier transform and chaos *Opt. Laser Eng.* **46** 117–23
[10] Lang J, Tao R and Wang Y 2010 Image encryption based on the multiple-parameter discrete fractional Fourier transform and chaos function *Opt. Commun.* **283** 2092–6

[11] Zhou N, Wang Y, Gong L, He H and Wu J 2011 Novel single channel color image encryption based on chaos and fractional Fourier transform *Opt. Commun.* **284** 2789–96

[12] Ye R 2011 A novel chaos-based image encryption scheme with an efficient permutation-diffusion mechanism *Opt. Commun.* **284** 5290–8

[13] Wang X, Zhao J and Liu H 2012 A new image encryption algorithm based on chaos *Opt. Commun.* **285** 562–6

[14] Fu C, Chen J J, Zou H, Meng W H, Zhan Y F and Yu Y W 2012 A chaos-based digital image encryption scheme with an improved diffusion strategy *Opt. Express* **20** 2363–78

[15] Liu H and Nan H 2013 Color image security system using chaos-based cyclic shift and multiple-order discrete fractional cosine transform *Opt. Laser Technol.* **50** 1–7

[16] Liu J, Jin H, Ma L and Jin W 2013 Optical color image encryption based on computer generated hologram and chaotic theory *Opt. Commun.* **307** 76–9

[17] Elshamy A M, Rashed A N Z, Mohamed A E N A, Faragalla O S, Mu Y, Alshebeili S A and Samie F E A E 2013 Optical image encryption based on chaotic Baker map and double random phase encoding *J. Lightwave Technol.* **31** 2533–9

[18] Norouzi B, Seyedzadeh S M, Mirzakuchaki S and Mosavi M R 2015 A novel image encryption based on row-column, masking and main diffusion processes with hyper chaos *Multimed. Tools Appl.* **74** 781–811

[19] Liu X, Mei W and Du H 2016 Simultaneous image compression, fusion and encryption algorithm based on compressive sensing and chaos *Opt. Commun.* **366** 22–32

[20] Wang Z, Lv X, Wang H, Hou C, Gong Q and Qin Y 2016 Hierarchical multiple binary image encryption based on a chaos and phase retrieval algorithm in the Fresnel domain *Laser Phys. Lett.* **13** 036201

[21] Chatterjee M R, Mohamed A and Almehmadi F S 2018 Secure free-space communication, turbulence mitigation, and other applications using acousto-optic chaos *Appl. Opt.* **57** C1–13

[22] Kumar A and Nishchal N K 2019 Quick response code and interference-based optical asymmetric cryptosystem *J. Inform. Sec. Appl.* **45** 35–41

IOP Publishing

Optical Cryptosystems

Naveen K Nishchal

Chapter 10

Optical asymmetric cryptosystems

10.1 Introduction

The optical techniques of information security have attracted the increased attention of researchers after the publication of the first pioneering work called DRPE. Since then a large number of techniques have been reported, expanding the scope with the use of different transforms (Fourier, Fresnel, fractional Fourier, gyrator, wavelet, cosine, Mellin). Analyzing the optical encryption schemes reported in literature, it is inferred that most of them can be categorized as symmetric cryptosystems, in which the security keys used for encryption and decryption are identical. Therefore, under an environment of network security, a symmetric cryptosystem may suffer from problems in key management and delivery. Also, in several studies it has been proved that the basic DRPE scheme is vulnerable. This is due to the inherent linearity in the DRPE method. To address this weakness, optical asymmetric cryptosystems have been reported [1–12]. An asymmetric encryption scheme is similar to symmetric encryption scheme except an asymmetry in the key design. Figure 10.1 depicts the classification of optical cryptosystems. The asymmetric nature was achieved through the amplitude- and phase-truncation method. This approach was introduced with the motive that it will break the involved linearity in the conventional DRPE technique.

An optical asymmetric cryptosystem (OAC) works on a similar principle to that of public key cryptography. A message encoded with one key should be decoded with another key. One key is treated as a private key while the other one is said to be the public key. An important point to note is that the public and private keys should be independent of the plaintext otherwise it will be like a one-time pad. Following the key design rule of asymmetric cryptography, the scheme should meet the following conditions [1].

(i) A pair of keys, a public key (encryption key) and a private key (decryption key) should be created.

doi:10.1088/978-0-7503-2220-1ch10

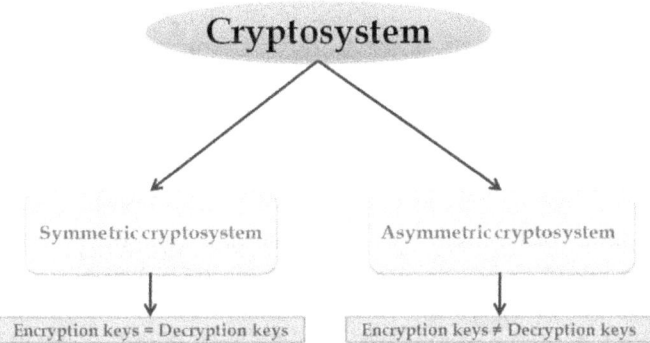

Figure 10.1. Classification of a cryptosystem.

(ii) Having known the plaintext and the encryption key, the ciphertext should be easily generated, following some encryption procedure.

(iii) It should be easy to retrieve the plaintext with the use of the correct private key following the reverse of the encryption procedure.

(iv) If an attacker gets access to the encryption key, it should be hard to generate the decryption key.

(v) If an attacker somehow accesses the encryption key and ciphertext, it should still be hard to retrieve the plaintext.

In an OAC, an input image/data is encrypted with the use of two RPMs placed in the object and frequency plane, respectively. An input image modulated with an RPM is Fourier transformed and the obtained spectrum, which is a complex function, is separated into amplitude and phase-only parts. This process has been referred to as amplitude- and phase-truncation. The amplitude-truncated part, which retains phase only information, is defined as the first decryption key. The phase-truncated part, which has amplitude information only, is further modulated with another RPM and one more Fourier transformation is obtained. Thus the obtained function is again a complex function. It is further amplitude- and phase-truncated. The amplitude-truncated part is retained as a second decryption key and the phase-truncated function is referred to as the encrypted image. Thus, in this DRPE scheme, two different RPMs are used and during the encryption process two decryption keys (phase-only functions) are generated. For successful decryption, only the decryption keys are required, the encryption keys (RPMs) are no longer needed. Thus, the storage of RPMs used for encoding information is not at all required for transmission. But for the retrieval of a secured message, the decryption keys are required. It is for this reason the scheme is referred to as an asymmetric cryptosystem. This process of amplitude- and phase-truncation is known as phase-truncated Fourier transform (PTFT). It breaks the linearity in the DRPE system and has been proved resistant to various attacks. In this scheme, the encryption process cannot be reversed because the phase truncation leads to a one-way function [2].

The principle of PTFT has been extended to other optical transforms in order to combine the associated features of the transforms for achieving higher security.

10.2 Asymmetric cryptosystems

Deriving encryption keys with optical parameters, an asymmetric cryptography based on wavefront sensing has been reported. In this proposal, optical parameters, such as wavelength, focal length, and their combination contributed to public key and a regular point array formed by microlenslet array was treated as the private key. The ciphertext was generated with the encoded wavefront and was represented with an irregular array. The process created a trapdoor one-way function [1]. Extending the concept of generating asymmetric keys further, an asymmetric cryptosystem based on twice PTFT was proposed [2]. The phase truncation approach generated a real-valued ciphertext, a stationary white noise. The generated two decryption keys were referred to as a universal key and special key. Figure 10.2 depicts the schematic of an asymmetric cryptosystem based on amplitude- and phase-truncation. Although the approach offered high-level security but suffered with key distribution problem. In a further study, the PTFT-based scheme was proved to be vulnerable to the specific attack [3]. The analysis was based on iterative Fourier transforms when two RPMs were used as public keys to encode different plaintexts [3]. In order to make the system resistant to specific attack and enhance the security, two methods were suggested [4]. The first suggested technique was to extend the PTFT-based cryptosystem to anamorphic FRT domain and the second

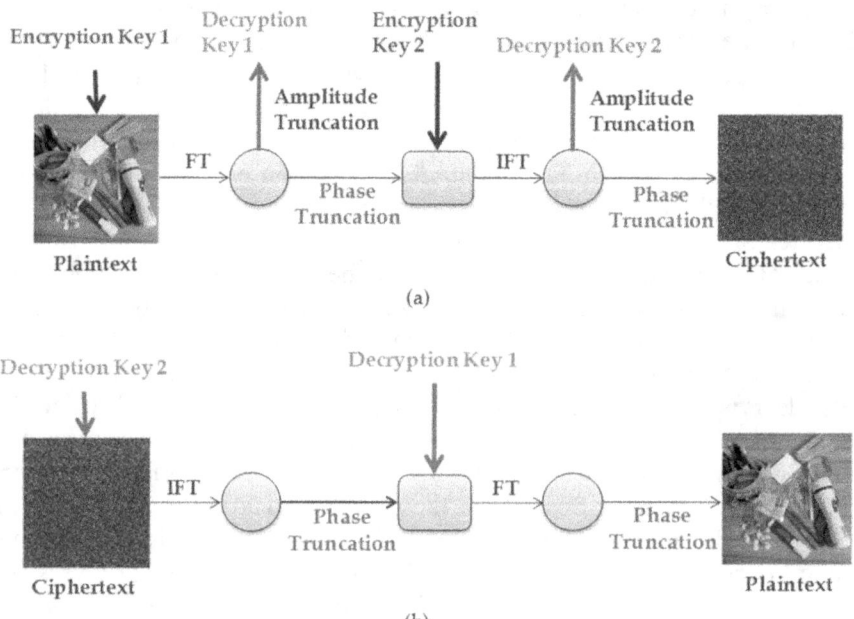

Figure 10.2. Schematic diagram for an amplitude- and phase-truncation-based asymmetric cryptosystem. (a) Encryption process and (b) decryption process. FT: Fourier transform and IFT: inverse Fourier transform.

suggestion was to introduce an undercover amplitude modulator. However, the suggested methods still suffered with the problem of key distribution and management.

The PTFT-based scheme has been applied for color image encryption employing interference of polarized wavefronts through polarization selective DOE and wavelength-dependent SPM [5]. The PTFT scheme has been implemented in the FRT domain. The PTFT-based method was improved with the use of spherical wave illumination rather than a uniform plane wave [6]. The scheme survived the special attack and in this case, the parameters of the spherical wave provided additional security, which were not made public. In order to redesign the public and private keys so that ciphertext depends upon both the keys in a complicated manner, an asymmetric cryptosystem was proposed. The method used the Yang-Gu mixture amplitude-phase retrieval algorithm [7]. This idea was commented with the advice that an easy and efficient way could be based on the PTFT approach to achieve the encryption-decryption [8]. In addition, it was also commented that the generation of private keys is plaintext dependent therefore, it violates the design rule. In a reply to the comment, it was argued that though there exists mutual interdependence between the private and public keys, they are connected by the trapdoor one-way function. However, it is difficult to generate a decryption key from the encryption key without knowing the trapdoor information [9].

An image encryption and authentication technique based on asymmetric keys generated by a phase retrieval algorithm and phase-truncation approach has been proposed [10]. Multiple input images bonded with multiple RPMs were Fourier transformed and the Fourier spectra were amplitude- and phase-truncated. The phase-truncated functions were multiplexed into a single random intensity image using a phase retrieval algorithm. The proposed method required less storage as only one encrypted image was required for the verification of multiple secured images. This idea could be extended for other security applications, such as image watermarking and hiding. An optical asymmetric cryptosystem has been reported that used the principle of elliptical polarized light linear truncation [11]. The method was characterized as pixel-to-pixel confusion-based without using a Fourier lens and diffusion operation. For polarized light reconstruction, the Jones matrix formalism was followed. For analyzing the key sensitivity and fault tolerance, a QR code was used.

In a different approach than the PTFT-based method, an asymmetric cryptosystem based on coherent superposition and equal modulus decomposition (EMD) has been reported [12]. The EMD based on interference was applied for creating an effective trapdoor one-way function and overcoming the specific attack problem. Also, this method does not have a silhouette problem. A fully-phase image encryption under the PTFT framework in a gyrator wavelet transform domain has been reported [13]. The gyrator wavelet transform constituted four additional parameters: an order of gyrator transform, type and level of mother wavelet, position of different frequency bands. These parameters served as keys in addition to the RPMs. Also, encryption and compression can both be achieved simultaneously. In a further study, an encryption technique based on equal modulus decomposition in a gyrator domain has been reported, in which the RPM was suppressed [14].

The input image was subjected to multiple images sampling at three different distances and the resultant was decomposed into two equal modulus parts. This cryptosystem offered a lesser number of constraints that could be known to an attacker and became specific attack resistant.

In order to manipulate complex-valued functions, a technique using the triplet of functions has been reported, which avoids using a holographic setup to record phase [15]. The approach was based on PTFT in the gyrator domain, in which the complex function is split into two matrices containing real and imaginary parts. These two matrices and a random distribution function were acted upon by one of the functions in the triplet. For decryption, the other two functions in the triplet help retrieve the complex-valued function. In a recent study, an asymmetric encryption method has been proposed for colour images using fractional Hartley transform [16].

10.3 Phase retrieval

A phase retrieval algorithm is a method for obtaining phase information from Fourier amplitude [17–25] and it is used in many fields, including cryptographic techniques. When recording an image with optical devices such as CCD cameras, only the photons, i.e. amplitude information, are recorded because phase value of light cannot be measured directly. In this case, the amount of information (phase) of the image is largely lost. Phase information carries more information than the magnitude of the image. The first phase retrieval algorithm was introduced by Gerchberg and Saxton in 1972, called the Gerchberg–Saxton (GS) algorithm [17]. Later on, many phase retrieval algorithms have been developed such as Yang-Gu, error reduction (ER), hybrid-input output (HIO), Wirtinger flow, and prDeep. The Wirtinger flow algorithm has two components: a careful initialization based on a spectral method and a series of updates refining the initial estimate by iteratively applying update rules, which have low computational complexity like a gradient descent scheme [19]. prDeep is a phase retrieval algorithm based on a convolutional neural network focusing on plug-and-play priors and deep learning priors [20]. Recently, a modified Fienup algorithm has been proposed that uses a controlled sparsity-enhancing step. It is applied such that in every iteration the complexity of the resulting guess solution is made close to the complexity parameter [21].

While the phase information can be obtained more smoothly with the developed algorithms, the backward access time to the phase is prolonged. The slow-running phase retrieval algorithms are not preferred for cryptographic techniques where processing time is vital. For this reason, the ER algorithm, which can restore the phase in a short time, is preferred.

In 2D phase retrieval-based encryption algorithms, when the input image is encoded with phase masks, phase masks are updated iteratively using the phase retrieval algorithm according to the correlation coefficient (CC) of the image being encrypted. When the CC value reaches the desired level, the phase masks are not updated anymore and the encryption operation is complete. In order to enhance the security, phase retrieval-based asymmetric encryption algorithms have been developed. Although the resistance to attacks is increased, the data that is encrypted can still be solved with the known-plaintext, ciphertext-only, and chosen-plaintext attacks. For this reason, in order to increase resistance to attacks, 2D phase retrieval-based encryption methods have been upgraded to a 3D version, in which the input image is divided into particles and then these particles are distributed into 3D space using RPMs.

10.3.1 Phase retrieval for security

Optical encryption techniques essentially use different aspects of the electromagnetic beam to hide information. In all such cryptosystems, the phase of the beam carrying the hidden information plays an important role. Since most of the cryptosystems are based on free space propagation which can be described by Fourier, Fresnel, or fractional Fourier transforms, retrieving the phase information using iterative or non-iterative phase retrieval algorithms find relevance in such systems. The phase retrieval technique has found applications in encryption, decryption and the cryptanalysis in optical cryptography. This section discusses the role of phase retrieval methods in context to the optical cryptosystems.

10.3.2 Phase retrieval: mathematical formulation

Let a 2D real space distribution function of an object be given by $g(x,y)$. The 2D Fourier transform of $g(x,y)$ is expressed by

$$G(f_x, f_y) = \int_{-\infty}^{\infty} \int g(x, y) \exp\left[-i2\pi(f_x x + f_y y)\right] \mathrm{d}x \mathrm{d}y \qquad (10.1)$$

The Fourier phase retrieval problem is to retrieve $g(x,y)$ when only magnitude of the Fourier spectrum, i.e. $|G(f_x, f_y)|$, is known.

The most common class of phase retrieval algorithm is the iterative ones based on alternate projections. The widely used GS algorithm falls into this category [17]. In this approach, two magnitude measurements for two different planes are known, such that the planes are related by Fourier transformation. The algorithm proceeds by iteratively going back and forth from the imaging plane to the Fourier domain and updating with the known information (magnitude measurements) in each step. Later, the GS algorithm was extended by incorporating different constraints leading to the ER algorithm and the HIO algorithm [18].

10.3.3 Modified GS algorithm for image multiplexing and encryption

Phase retrieval algorithms have been used in different contexts in optical cryptosystems. To aid the general understanding as to how the phase retrieval algorithms find a place in encryption setups, here, one of the approaches, wherein the modified GS algorithm (MGSA) has been used for multiple images encryption, is discussed [26]. The domain used for encryption is the Fresnel domain. The conventional GS algorithm is modified to evaluate a phase-only function (POF) such that the Fresnel transform of the POF gives the plaintext image. In this iterative procedure, the amplitude in the object domain is constrained to unity. Figure 10.3 shows the

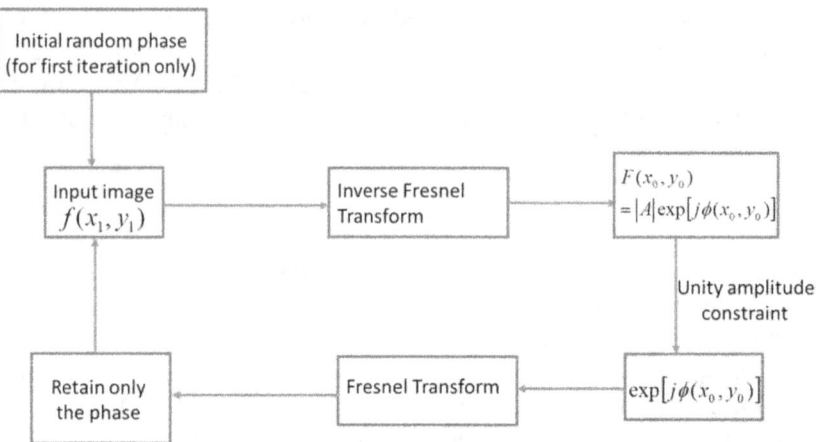

Figure 10.3. Flowchart depicting the MGSA used to construct the POF.

flowchart for the MGSA used in the encryption scheme. Let $f(x_1, y_1)$ represent an input plaintext image. To initiate the algorithm, the image is multiplied with an RPM, and subsequently subjected to the inverse FrT to obtain an intermediate complex function, $F(x_0, y_0)$. The phase of this intermediate function, denoted by $\varphi_g(x_0, y_0)$, is retained as it is and the amplitude is constrained to take the value of unity. The phase is then Fresnel transformed where the amplitude is updated with the plaintext information. This process is repeated until a predefined correlation is achieved between the approximated amplitude $f'(x_1, y_1)$ obtained after Fresnel transform and the original image, $f(x_1, y_1)$.

For the case of multiple images encryption and multiplexing, different POFs need to be evaluated corresponding to different input images. The main advantage of this approach is that different wavelengths and propagation distances can be used for the Fresnel transformation for each of the images. Thus, a unique POF is generated with different parameters for each image. If there are $n = 1$ to N images to be multiplexed, and if different wavelengths are being used for the generation of the POF, it is said to be wavelength multiplexed. Mathematically, the relation between the POF and the input image can be stated as

$$\Im_\lambda^d \{\exp[j\varphi_{\lambda_n}(x_0, y_0)]; \lambda_n; d\} = f_n^\lambda(x_1, y_1)\exp[j\psi_{\lambda_n}(x_1, y_1)] \qquad (10.2)$$

The subscript λ_n denotes the wavelength used for the nth image. When different propagation distances d_n are used for POF generation, it is said to be position multiplexed. The relation between POF and the input image is then stated as

$$\Im_\lambda^{d_n} \{\exp[i\varphi_{d_n}(x_0, y_0)]; \lambda; d_n\} = f_n^d(x_1, y_1)\exp[i\psi_{d_n}(x_1, y_1)] \qquad (10.3)$$

Figure 10.4 shows the numerical simulation result for a POF generated using the MGSA.

Figure 10.4. Numerical simulation results showing (a) the input image used in MGSA, (b) phase value distribution of the POF, and (c) the image retrieved by Fresnel transform of the POF.

The N wavelengths (or positions) multiplexed POFs are combined together by summation to obtain a single POF. The nth original plaintext image $f_n(x_1, y_1)$ can be recovered by the Fresnel transform of the final POF with the appropriate wavelength λ and propagation distance d. However, such a recovered image would contain cross-talk terms. To overcome this, each $f_n(x_1, y_1)$ is spatially translated to different positions by appropriate phase modulation of the constituent POF. The aim of this phase modulation is to project each of the multiplexed images at different spatial co-ordinates in the same plane on decryption. For this, the POF is modified as follows (the following equation considers wavelength multiplexing but holds for position multiplexing too):

$$\varphi'_{\lambda_n}(x_0, y_0) = \varphi_{\lambda_n}(x_0, y_0) + \frac{2\pi(\mu_n x_0 + \nu_n y_0)}{\lambda_n d} \tag{10.4}$$

The variables μ_n and ν_n denote the respective amount of shifts for the nth image along the x_1 and y_1 directions. The Fresnel transform of the POF in this case would yield

$$\mathfrak{I}^d_\lambda \left\{ \exp\left[j\varphi'_{\lambda_n}(x_0, y_0) \right]; \lambda_n; d \right\} - f_n^\lambda(x_1 - \mu_n, y_1 - \nu_n) \exp\left[j\theta(x_1, y_1) \right] \tag{10.5}$$

Here, $\theta(x_1, y_1)$ is the accompanied phase term. The cross-talk can be minimized by the appropriate choice of μ_n and ν_n. The final POF is generated from the summation and normalization of POFs corresponding to each image:

$$\varphi_S^\lambda(x_0, y_0) = \arg\left\{ \frac{\sum\limits_{n=1}^{N} \exp\left[j\varphi'_{\lambda_n}(x_0, y_0) \right]}{\left| \sum\limits_{n-1} \exp\left[j\varphi'_{\lambda_n}(x_0, y_0) \right] \right|} \right\} \tag{10.6}$$

Here, arg refers to the argument operator. Position multiplexing can be represented in the same manner.

Hence, the scheme uses the MGSA for the generation of the POF, which is further used for encryption and multiplexing of multiple images (figure 10.5). The generation of POF has seen a wide application in phase retrieval-based optical cryptosystems.

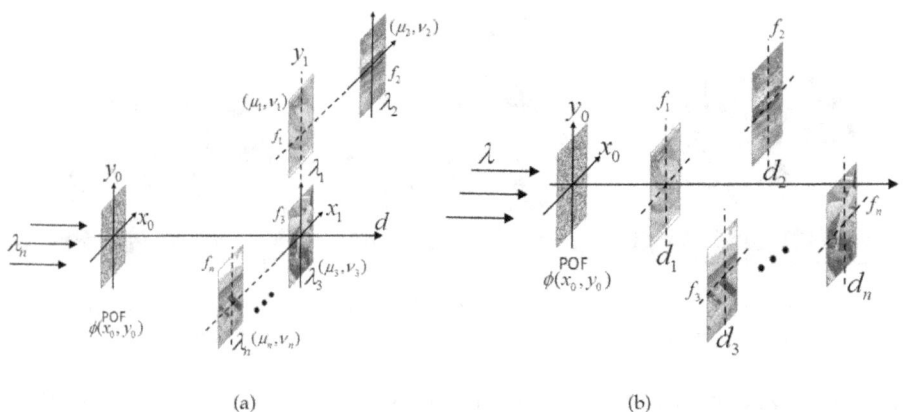

Figure 10.5. (a) Wavelength demultiplexing and (b) position demultiplexing in the optical cryptosystem using POF generated by MGSA in the Fresnel domain.

The early work incorporating the phase retrieval algorithms involved security systems based on the 4f optical processor, such as the VanderLugt architecture [27]. Such a system used a phase function in the frequency plane which was obtained using a modified phase retrieval algorithm. This kind of scheme gave different off-shoots of optical encryption methods using a phase retrieval algorithm. Two or more phase masks have been shown to be iteratively retrieved by implementing a multiple-image phase retrieval algorithm to get the target image [28]. The phase retrieval algorithm has been extended to other optical transformations such as FRT and FrT domains. The use of FRT offers additional keys in the form of the fractional order. In one such approach, the input image is bonded with an RPM and subjected to FRT to obtain the intensity in this domain. This process was repeated with a different RPM and a different order of FRT to obtain a second intensity measurement. The decryption comprised use of these intensity measurements and the RPMs in a phase retrieval algorithm to obtain the plaintext [29]. Asymmetric cryptosystems have been established, which uses a GS algorithm twice for obtaining two levels of encryption. The method allows two asymmetric keys to be generated which can then be used in the conventional DRPE system. The scheme showed immunity against various attacks [30].

10.4 Photon counting imaging

To prevent various applicable attacks on schemes based under the DRPE framework, photon counting imaging (PCI) has been combined with an encryption system [31–37]. Use of PCI to the encrypted image generates a photon-limited image, thus introducing an additional layer of security. The PCI technique controls the number of photons that arrive at a pixel through a stochastic Poisson process. Applying the PCI technique makes the image/data sparsely represented. Therefore, the output images do not resemble original images and cannot be visually distinguished. This important property has been proposed to safeguard DRPE-based encryption schemes from unauthorized attacks. Mostly, this scheme does not intend to visualize input information, but to authenticate the original image/data. However, it has been reported that PCI-based schemes can also be used for revealing the original information. The 3D photon counting DRPE scheme has been reported, in which 3D information was revealed using passive integral imaging [33]. In this study, it was claimed that a 3D photon counting DRPE can encrypt a 3D scene and provide higher security and authentications. Also, 3D imaging allowed verification of the 3D object at different depths. A scheme based on PCI and DRPE has been reported for full phase encoding [35]. PCI was used during the encryption process for creating sparse images, which were further used for authentication application. A photon-counting polarimetric technique for encryption and verification has been reported [34]. Generating random polarized vector keys using a Mach–Zehnder configuration was developed, in which the polarization information of the encrypted signal was retrieved by computing the Stokes parameters. The photon-counting model was used during the encryption process to provide sparse data and nonlinear transformation in order to achieve an enhanced level of security. Any authorized user having access of the polarization keys and the optical design variables can retrieve and validate the photon-limited plaintext.

The PCI technique is modeled as a Poisson distribution [31–37]. Suppose the number of photons counts in entire space is N_p. The probability of counting l_i photons at pixel i has been shown to be Poisson distributed [31],

$$P(l_i, \lambda_i) = \frac{[\lambda_i]^{l_i} \exp(-\lambda_i)}{l_i!}, \qquad l_i = 0, 1, 2 \dots \tag{10.7}$$

where l_i is the number of photons that arrived at pixel i. The Poisson parameter, λ_i is obtained with the help of normalized irradiance at pixel i,

$$\lambda_i = N_p g(i) \tag{10.8}$$

Suppose M is the total number of pixels in the entire image. Now, the normalized irradiance, $g(i)$ is defined at pixel i, such that [29],

$$\sum_{i=1}^{M} g(i) = 1 \tag{10.9}$$

In PCI, sparse content is inversely proportional to the number of photons *i.e.*, when less photons arrive at the scene, the image will be sparser. The pixels must receive at least one photon count to apply the PCI technique in the photon-limited encryption function [31].

10.5 PCI and phase-truncated FrT-based asymmetric encryption

The main intention of integration of DRPE with PCI was invisible authentication. The decrypted image so obtained cannot be easily seen by attackers. However, if a sufficient number of photons is used during encryption then a good quality original image can be retrieved. When the retrieved image becomes visible, it means there is a good correlation between the encrypted image and the photon limited encrypted image. This will make the PCI-DRPE system vulnerable to various types of attacks

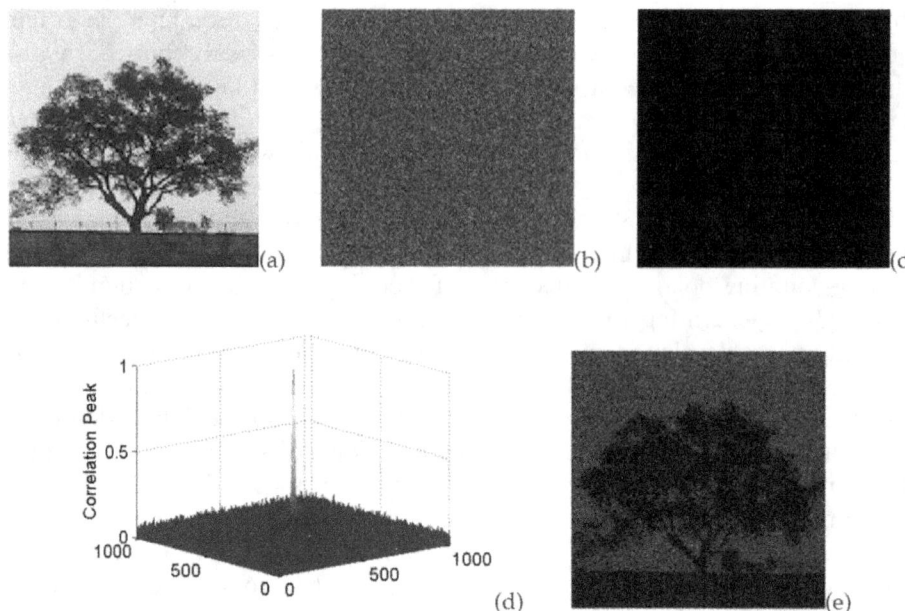

Figure 10.6. Simulation results: (a) the input image, (b) the phase-truncated Fresnel transform-based encrypted image, (c) the PCI phase-truncated Fresnel transform-based encrypted image with 10^3 photons, (d) the correlation peak when the retrieved image is matched with the corresponding original image, and (e) the decrypted image obtained after using all the correct keys.

effective on the DRPE technique. It was suggested that by integrating the PCI with the PTFT approach, this problem can be resolved successfully. Conceptualizing the suggestion, an optical asymmetric cryptosystem based on PCI and phase-truncated FrT has been reported [37].

The simulation results based on this technique has been presented in figures 10.6(a)–(e). An image, as shown in figure 10.6(a), to be encoded or authenticated is converted into a real-valued noisy image (figure 10.6(b)) using the phase truncation approach in the FrT domain. The obtained noisy image is further encoded using the PCI technique using 10^3 photons, as shown in figure 10.6(c). For the retrieval of the original image for authentication, the reverse process of the phase truncation approach was followed and a nonlinear correlation filter was used. The auto-correlation peak when the retrieved image is matched with the corresponding original image is shown in figure 10.6(d), which confirms authentication of the encrypted data. Figure 10.6(e) shows the retrieved image after using all the correct keys but with only 10^3 photons. It can be inferred that with the photon-limited image, authentication is possible but complete reconstruction of the original image is not possible. For successful content retrieval, the use of sufficient numbers of photons is recommended. The retrieved image obtained after using all the correct keys and with 10^6 photons is the same as what is shown in figure 10.6(a). Therefore, the technique can be used for complete information retrieval as well as for authentication, depending upon the available number of photons.

An important issue related with the PTFT-based asymmetric cryptosystem is that the encrypted information can be disclosed only with the use of the first decryption key. This key contains the phase information of image spectrum and therefore can get the silhouette of an image in the Fourier domain. However, if the encryption is carried out in other domains like FRT or FrT domains, the system will have additional keys. So, information disclosure will not be possible without knowledge of these additional keys.

Asymmetric techniques are preferred over symmetric schemes of image encryp tion due to obvious reasons. But asymmetric schemes have been proved vulnerable to specific attack, which further excited the research community to come up with advanced techniques to counter the weaknesses. The optical asymmetric cryptosystems still need to be strengthened with more emphasis on enhancing its capability to withstand attacks and make it suitable for a practical cryptosystem.

MATLAB code for phase retrieval

```
%%Matlab code for phase-only function generation using
modified Gerchberg-Saxton algorithm (MGSA)
%%The code uses the Fourier transform for the MGSA. For
Fresnel domain, the Fourier transform can be easily substituted
with the Fresnel transform function
g0 = imread('z256tree.png');
g0 = double(g0(:,:,1));
g0 = g0./max(max(g0));
figure;imagesc(abs(g0));colormap(gray);axis off; axis equal;
title('input image');
phase1=rand(256);
f1=(g0/(sqrt(2))).*exp(1i*pi*phase1);
%iteration
n = input('enter the number of iteration = ');

for j=1:n
 F=fftshift(ifft2(ifftshift(f1)));
 %Fourier transforming the quantity ER
 phase2=angle(F);
 F=1.*exp(1i*phase2);%POF
 f1=fftshift(fft2(ifftshift(F)));
 f1=(g0/(sqrt(2))).*exp(1i*angle(f1));
end
g0_ret=fftshift(fft2(ifftshift(F)));
figure;imagesc(abs(phase2));colormap(gray);
title('Phase value distribution of POF');
figure;imagesc(abs(g0_ret));colormap(gray);
title('Input image retrieved after Fourier transform of POF');
```

References

[1] Peng X, Wie H and Zhang P 2006 Asymmetric cryptography based on wavefront sensing *Opt. Lett.* **31** 3579–681

[2] Qin W and Peng X 2010 Asymmetric cryptosystem based on phase-truncated Fourier transforms *Opt. Lett.* **35** 118–20

[3] Wang X and Zhao D 2012 A special attack on the asymmetric cryptosystem based on phase-truncated Fourier transforms *Opt. Commun.* **285** 1078–81

[4] Wang X and Zhao D 2011 Security enhancement of a phase truncation based image encryption algorithm *Appl. Opt.* **50** 6645–51

[5] Rajput S K and Nishchal N K 2012 Asymmetric color cryptosystem that uses polarization selective diffractive optical element and structured phase mask *Appl. Opt.* **51** 5377–86

[6] Ding X, Deng X, Song K and Chen G 2013 Security improvement for asymmetric cryptosystem based on spherical wave illumination *Appl. Opt.* **52** 467–73

[7] Liu W, Liu Z and Liu S 2013 Asymmetric cryptosystem using random binary phase modulation based on mixture retrieval type of Yang-Gu algorithm *Opt. Lett.* **38** 1651–3

[8] He W, Meng X and Peng X 2013 Asymmetric cryptosystem using random binary phase modulation based on mixture retrieval type of Yang-Gu algorithm: comment *Opt. Lett.* **38** 4044

[9] Liu W, Liu Z and Liu S 2013 Asymmetric cryptosystem using random binary phase modulation based on mixture retrieval type of Yang-Gu algorithm: reply *Opt. Lett.* **38** 4045

[10] Rajput S K and Nishchal N K 2014 An optical encryption and authentication scheme using asymmetric keys *J. Opt. Soc. Am.* A **31** 1233–8

[11] Lin C, Shen X, Wang Z and Zhao C 2014 Optical asymmetric cryptography based on elliptical polarized light linear truncation and a numerical reconstruction technique *Appl. Opt.* **53** 3920–8

[12] Cai J, Shen X, Lei M, Lin C and Dou S 2015 Asymmetric optical cryptosystem based on coherent superposition and equal modulus decomposition *Opt. Lett.* **40** 475–8

[13] Mehra I and Nishchal N K 2015 Optical asymmetric image encryption using gyrator wavelet transform *Opt. Commun.* **354** 344–52

[14] Fatima A, Mehra I and Nishchal N K 2016 Optical image encryption using equal modulus decomposition and multiple diffractive imaging *J. Opt.* **18** 085701

[15] Yatish, Fatima A and Nishchal N K 2018 Optical image encryption using triplet of functions *Opt. Eng.* **57** 033103

[16] Yadav A K, Singh P, Saini I and Singh K 2019 Asymmetric encryption algorithm for colour images based on fractional Hartley transform *J. Mod. Opt.* **66** 629–42

[17] Gerchberg R W and Saxton W O 1972 A practical algorithm for the determination of phase from image and diffraction plane pictures *Optik* **35** 237–50

[18] Fienup C and Dainty J 1987 Phase retrieval and image reconstruction for astronomy *Im. Recov.: Theor Appl.* **231** 275

[19] Candes E J, Li X and Soltanolkotabi M 2015 Phase retrieval via Wirtinger flow: theory and algorithms *IEEE Trans. Inform. Th.* **61** 1985–2007

[20] Metzler C, Schniter P, Veeraraghavan A and Baraniuk R 2018 prDeep: robust phase retrieval with a flexible deep network *35th Intl Confer. Machine Learning (JMLR.org)* pp 3498–507

[21] Butola M, Rajora S and Khare K 2019 Phase retrieval with complexity guidance *J. Opt. Soc. Am.* A **36** 202–11

[22] Shechtman Y, Eldar Y C, Cohen O, Chapman H N, Miao J and Segev M 2015 Phase retrieval with application to optical imaging: a contemporary overview *IEEE Sig. Process. Mag.* **32** 87–109

[23] Millane R P 1990 Phase retrieval in crystallography and optics *J. Opt. Soc. Am.* A **7** 394–411

[24] Faulner H M L and Rodenburg J M 2004 Movable aperture lensless transmission microscopy: a novel phase retrieval algorithm *Phys. Rev. Lett.* **93** 023903

[25] Chen W, Chen X and Sheppard C J 2012 Optical image encryption based on phase retrieval combined with three-dimensional particle-like distribution *J. Opt.* **14** 075402

[26] Hwang H E, Chang H T and Lie W N 2009 Multiple-image encryption and multiplexing using a modified Gerchberg–Saxton algorithm and phase modulation in Fresnel transform domain *Opt. Lett.* **34** 3917–9

[27] Wang R K, Watson I A and Chatwin C 1996 Random phase encoding for optical security *Opt. Eng.* **35** 2464–9

[28] Chang H T, Lu W C and Kuo C J 2002 Multiple-phase retrieval for optical security systems by use of random-phase encoding *Appl. Opt.* **41** 4825–34

[29] Hennelly B and Sheridan J T 2003 Fractional Fourier transform-based image encryption: phase retrieval algorithm *Opt. Commun.* **226** 61–80

[30] Rajput S K and Nishchal N K 2014 Fresnel domain nonlinear image encryption scheme based on Gerchberg-Saxton phase retrieval algorithm *Appl. Opt.* **53** 418–25

[31] Perez-Cabre E, Abril H C, Millan M S and Javidi B 2012 Photon-counting double-random-phase encoding for secure image verification and retrieval *J. Opt.* **14** 094001

[32] Chen W, Chen X, Stern A and Javidi B 2013 Phase-modulated optical system with sparse representation for information encoding and authentication *IEEE Photon. J* **5** 6900113

[33] Cho M and Javidi B 2013 Three-dimensional photon counting double-random-phase encryption *Opt. Lett.* **38** 3198–201

[34] Markman A and Javidi B 2014 Full-phase photon-counting double-random-phase encryption *J. Opt. Soc. Am.* A **31** 394–403

[35] Maluenda D, Carnicer A, Martinez-Herrero R, Juvells I and Javidi B 2015 Optical encryption using photon-counting polarimetric imaging *Opt. Express* **23** 655–66

[36] Rajput S K, Kumar D and Nishchal N K 2015 Optical encryption system based on phase mask multiplexing and photon counting imaging for multiple image authentication and digital hologram security *Appl. Opt.* **54** 1657–66

[37] Rajput S K and Nishchal N K 2017 Optical asymmetric cryptosystem based on photon counting and phase-truncated Fresnel transforms *J. Mod. Opt.* **64** 878–86

IOP Publishing

Optical Cryptosystems

Naveen K Nishchal

Chapter 11

Attacks on optical security schemes

11.1 Introduction

A good encryption system should be robust against all kinds of applicable attacks. The security of encrypted information solely depends on the security of the secret encryption keys. In this regard, it is important to mention Kerckhoffs's principle, which states that 'A cryptographic system should be secure even if everything about the system, except the key, is public knowledge'. This is one of the basic and relevant principles of modern cryptography, which was formulated in 1883 by a Dutch cryptographer Auguste Kerckhoffs [1].

Optical transformations such as Fourier transformation are linear, and the linearity leads to security leaks in DRPE and its derivative encryption systems. Cryptanalysis reveals that DRPE-based cryptosystems are vulnerable to some attacks, including ciphertext-only attack (COA), known-plaintext attack (KPA), and chosen-plaintext attack (CPA) [2–7]. In COAs, the interception is through phase retrieval procedure with single intensity measurement. The plaintext is recovered from its corresponding ciphertext via iterative phase retrieval algorithms such as the Gerchberg–Saxton (GS) algorithm and hybrid input–output (HIO) algorithm [3]. This kind of attack requires object-domain constraints and corresponding Fourier domain constraints. The latter constraints are provided by amplitude distribution in Fourier plane in the encryption system. The former constraints are provided by the zero-value data points in the input plane, or essentially by the free information capacity that are unoccupied by the input signal.

In KPAs, the encryption key is retrieved from one or more known plaintext-ciphertext pairs, and then the retrieved key is used to decrypt other ciphertexts encrypted with the same key. The encryption key is usually obtained in KPAs via, firstly, phase retrieval algorithms such as the GS or HIO algorithm, secondly, heuristic iterative algorithms such as the simulated annealing (SA) algorithm, and thirdly, solving linear equations. Besides the ciphertext, the corresponding original image is also known to interceptors. The SA algorithm runs iteratively to search for

doi:10.1088/978-0-7503-2220-1ch11

the global optima of the cost function and is usually very time-consuming. The KPA based on the SA algorithm is essentially repeated to search for the correct key from the whole key space. Therefore, this attack does not depend on the encryption relationship between the ciphertext and original image but depends on the decryption relationship between the decryption result and ciphertext.

In CPAs, special plaintexts, such as Dirac delta function, are encrypted, and then the encryption key is obtained from the chosen plaintexts and their corresponding ciphertexts. The CPAs are usually based on impulse attack. To perform the CPA, the Dirac delta function is employed as the chosen plaintext, and is encrypted to retrieve the encryption key. The COA and KPA require less resources of the cryptosystem than CPA. The more the resource is accessed, the easier the cryptosystem is to cracked.

In the following sections, these attacks are discussed in detail.

11.2 Brute-force attack

Brute force attacks are the most elementary of all the attacks. It involves approximating the keys of a cryptosystem by random guesses and then using it to decipher the ciphertext. The process is repeated until the attacker finds the correct key. To elaborate this method, the case of DRPE has been discussed here. Assuming that two RPMs have a size of $\sqrt{N} \times \sqrt{N}$ pixels, if each pixel can take on L possible phase values then to retrieve both the keys, it requires a maximum of L^{2N} attempts. For a 16-phase level system, and a phase key of the size 100×100 pixels, this value becomes $16^{20\,000} \sim 10^{24\,000}$. This number is very large and increases rapidly as the number of pixels increase, making the brute force attack a lengthy process. However, there are different considerations that can make these attacks less computationally extensive. One of the simpler approaches can be to try every possible value for the second key used in the DRPE, ignoring the first key. This reduces the number of trials to L^N which is $\sim 10^{12\,000}$ for the case of the 16-phase level system. Similarly, approximate decryption can be achieved by simplifying the guessed key in each trial. For example, only binary keys can be tried for the second key, and the first key can be ignored. Binary key means using only two phase levels to approximate the phase keys. Hence, the number of combinations possible for the attempts to break the system in this case is reduced to 2^N instead of L^N. The binary phase approximation does not give exact decryption. However, it is sufficient to reveal the larger features of the plaintext [2]. Hence, it can be concluded that with different approximations, the brute force attack can be applied to the cryptosystems with reduced complexity.

11.3 Differential attack

It is a desirable feature of any cryptosystem that a minor change in the plaintext generates a prominent change in the ciphertext. This analysis is called the differential attack analysis and it makes sure that it is not easy to discern the plaintext by manipulating the ciphertext. The change in the ciphertext is measured by two standards, namely, the number of pixel change rate (NPCR) and the unified average change in intensity (UACI). If C_1 represents the initial ciphertext, C_2 represents the manipulated ciphertext and N represents the total number of pixels, then these parameters are defined as [4]

$$\text{NPCR} = \sum_{i,j} \frac{D(i,\ j)}{N} \times 100\% \tag{11.1}$$

where

$$D(i,\ j) = \begin{cases} 0, & \text{if } C_1(i,\ j) = C_2(i,\ j) \\ 1, & \text{if } C_1(i,\ j) \neq C_2(i,\ j) \end{cases} \tag{11.2}$$

$$\text{UACI} = \sum_{i,j} \frac{|C_1(i,\ j) - C_2(i,\ j)|}{2^n - 1} \times 100\% \tag{11.3}$$

Here, $(i,\ j)$ represents the pixel indices, and n represents the bit depth of the image. The NPCR measures the exact number of pixels that change when the plaintext is changed. Hence the higher the value of NPCR the better the robustness of the system. On the other hand, the UACI measures the average difference between corresponding pixel values of the two ciphertexts. In this case, lower values are better.

11.4 Known-plaintext attack

The known-plaintext attack aims at retrieving the encryption keys of a cryptosystem given that a pair of plaintext and the corresponding ciphertext are known to the attacker. Also, the encryption algorithm is assumed to be known to the attacker. In most of the optical cryptosystems, the KPA is formulated using phase retrieval algorithms, which involves evaluating phase information with the given intensity information in the Fourier domain. For example, the KPA on DRPE can be considered [5]. The encryption process can be written mathematically as

$$E(x,\ y) = \mathfrak{I}^{-1}\{\{\mathfrak{I}\{f(x,\ y)\exp[i2\pi R_1(x,\ y)]\}\}\exp[i2\pi R_2(u,\ v)]\} \tag{11.4}$$

Here, $f(x,y)$ and $E(x,y)$ represent the plaintext and ciphertext, respectively. The Fourier transform of the ciphertext can be written as

$$\mathfrak{I}\{E(x,\ y)\} = G(u,\ v) = \{\mathfrak{I}\{f(x,\ y)\exp[i2\pi R_1(x,\ y)]\}\}\exp[i2\pi R_2(u,\ v)] \tag{11.5}$$

Therefore,

$$|G(u,\ v)| = |\mathfrak{I}\{E(x,\ y)\}| \tag{11.6}$$

Also,

$$|G(u,\ v)| = |\mathfrak{I}\{f(x,\ y)\}\exp[i2\pi R_1(x,\ y)]| \tag{11.7}$$

Here, $f(x,y)$ is known to the attacker and $|G(u,\ v)|$ can be evaluated from the known ciphertext owing to equation (11.6). Hence, concentrating on equation (11.7), the problem reduces to knowing the amplitude information, $f(x,y)$ in the signal domain and amplitude information ($|G(u, v)|$) in the frequency domain, and the phase information $\exp[i2\pi R_1(x, y)]$ need to be evaluated. This is a classic phase retrieval problem and the phase can be recovered by using GS algorithm and related algorithms. Chapter 10 addresses the phase retrieval algorithms in detail. An iterative algorithm can be formulated that proceeds back and forth between the signal domain and the frequency domain, updating the obtained spectrum in each domain with the respective known quantities. The second encryption key can be obtained once the first decryption key $k_1 = \exp[-i2\pi R_1(x, y)]$ is obtained. This second phase key for decryption can be evaluated as

$$\exp[-i2\pi R_2(u, v)] = \frac{FT\{f(x, y)k_1\}}{G(u, v)} \tag{11.8}$$

This attack analysis showed that the DRPE was vulnerable to the KPA. The keys used in the system can be intercepted by any attacker if the plaintext and the corresponding ciphertext are known. The recovered keys can then be used for retrieving the plaintexts for other ciphertexts. Hence, for the DRPE system to be robust, it became important not to use one set of keys for different plaintext encryption. This led to a one-time pad system for robust DRPE implementation. Apart from the phase retrieval technique, a different known-plaintext attack can be formulated for breaching any cryptosystem [6–10]. For example, it has been shown that the SA heuristic algorithm can be used to estimate the keys, given that plaintext and corresponding ciphertext pair are known [6]. The asymmetric optical crypto-system has been shown vulnerable to the KPA, wherein the algorithm employed phase retrieval methods [10].

11.5 Chosen-plaintext attack

In a chosen-plaintext attack, the attacker has the knowledge of the encryption algorithm, and the ciphertexts can be obtained corresponding to random plaintexts of his/her choice. From this knowledge, the attacker tries to intercept further information about the encryption system such as the encryption keys. It has been shown that the DRPE is vulnerable to the CPA by choosing a series of impulse functions as plaintext and then obtaining the corresponding ciphertext to further retrieve the keys [11]. The vulnerability arises because of the fact that the second Fourier transform in the DRPE does not add anything to the security of the system. To understand this approach, equation (11.5) is rewritten as

$$G(u, v) = \{\mathfrak{F}\{f(x, y)\exp[i2\pi R_1(x, y)]\}\}\exp[i2\pi R_2(u, v)] \tag{11.9}$$

Now, let the initial chosen plaintext be an impulse function, $\delta(x - u, y - v)$. Considering the case $u = 0$ and $v = 0$, the plaintext becomes $\delta(x, y)$ which means that it has zero value at all the points except at the position (0,0). It can be easily seen from equation (11.9) that this yields a ciphertext which is (up to a constant phase factor), the second key, $\exp[i2\pi R_2(u, v)]$. If the non-zero pixel is not centred in the plaintext, i.e. if the plaintext is taken as an impulse function of the form

$\delta(x - u, y - v)$, then the ciphertext is the key $\exp[i2\pi R_2(u, v)]$ multiplied by a linearly varying phase. This linearly varying phase is dependent on the position of the non-zero pixel. Hence, it can be seen that it is possible to breach the system with the CPA. A similar approach has been shown for CPA on the DRPE system in the Fresnel domain. Modifications in this approach has been reported, which involves subtracting the ciphertexts corresponding to two plaintexts whose difference is a delta function [2]. It has been shown that the JTC-based encryption system suffers from vulnerability to the CPA [12]. An optical cryptosystem based on a scrambling pre-processing operation followed by DRPE has been shown to be vulnerable to the CPA based on this approach of using impulse functions [13].

11.6 Chosen-ciphertext attack

Chosen-ciphertext attack is one of the important attack analyses for any information security system. Resistance to this attack is widely considered as the minimum standard that a public key encryption system should adhere to. This attack is also known as the lunchtime attack or the midnight attack. In this attack, it is assumed that the attacker can access the decryption system. In the first stage, the attacker introduces different ciphertexts of his/her own choice in the decryption system and obtains the respective plaintexts. The second stage is the guessing stage, wherein, the attacker has no access to the decryption system and s/he is presented with a ciphertext corresponding to a randomly selected plaintext.

If the attacker is able to figure out the plaintext from this ciphertext, then the system is said to be vulnerable to the CPA. It has been shown that the DRPE can be vulnerable to this attack. The attacker repeatedly subjects the chosen ciphertexts to the decryption system to obtain the plaintext. Here, the chosen-ciphertext should be exclusively designed so as to retrieve as much information as possible about the keys [14]. Let this chosen ciphertext be of the following form,

$$E_c(x, y) = \frac{1}{2}\left[\exp(2\pi if_{x_1}x)\exp(2\pi if_{y_1}y) + \exp(2\pi if_{x_2}x)\exp(2\pi if_{y_2}y)\right] \quad (11.10)$$

Here, the subscript c denotes that the ciphertext is chosen by the attacker. To proceed with the attack, this chosen-ciphertext is subjected to the decryption path, where it is first Fourier transformed, which gives the following expression,

$$E_{1c}(u, v) = \frac{1}{2}\left[\delta(u - f_{x_1}, v - f_{y_1}) + \delta(u - f_{x_2}, v - f_{y_2})\right] \quad (11.11)$$

This expression contains two Dirac delta functions centred at points (f_{x_1}, f_{y_1}) and (f_{x_2}, f_{y_2}). If the key is represented by $\exp(-i2\pi R_3(u, v))$, then after passing through this key, followed by Fourier transform, the optical field expression in the output plane will be given by

$$f_c(x, y) = \Im[E_{1c}(u, v)\exp(i2\pi R_3(u, v))]$$

$$= 0.5\begin{Bmatrix} \exp[-i2\pi R_3(f_{x_1}, f_{y_1})]\exp(-i2\pi f_{x_1}x)\exp(-i2\pi f_{y_1}y) \\ + \exp[-i2\pi R_3(f_{x_2}, f_{y_2})]\exp(-i2\pi f_{x_2}x)\exp(-i2\pi f_{y_2}y) \end{Bmatrix} \quad (11.12)$$

The intensity profile then becomes

$$|f_c(x, y)| = 0.5 + 0.5\cos\Big[2\pi(f_{x_1} - f_{x_2})x + 2\pi(f_{y_1} - f_{y_2})y$$
$$+ 2\pi R_3(f_{x_2}, f_{y_2}) - 2\pi R_3(f_{x_1}, f_{y_1})\Big] \quad (11.13)$$

Since the frequencies $f_{x_1}, f_{x_2}, f_{y_1}$ and f_{y_2} are known, hence, the repeated application of this attack can be used to compute the values of $R_3(f_{x_2}, f_{y_2}) - R_3(f_{x_1}, f_{y_1})$. The security system can be breached in this way. To overcome such an impulse-based attack, a technique was proposed wherein the phase factor of the lenses used in the DRPE were multiplied with random-phase functions. This results in a modified Fourier transform, giving a random output for an impulse function instead of unity function. This makes the DRPE system safer from impulse attacks [15].

11.7 Specific attack

The specific attack was formulated for breaching the asymmetric optical cryptosystems. The asymmetric optical cryptosystems used decryption keys different than the encryption keys. The system was based on the phase-truncated Fourier transformation (PTFT). The decryption keys are generated during the encryption process, and are plaintext-dependent. The KPA aims to retrieve the keys of any cryptosystem. Hence, for a PTFT-based encryption system, where the keys change with the plaintext, any application of the KPA to intercept the keys becomes redundant. The generation of the ciphertext in this case can be mathematically stated as

$$E(u, v) = PT\{\Im^{-1}[\{PT[\Im(f(x, y) \times K_1(x, y))]\} \times K_2(m, n)]\} \quad (11.14)$$

Here, K_1 and K_2 denote the RPMs used as the encryption keys. The first specific attack to be proposed for this system aimed to retrieve the plaintext, given that the attacker has the knowledge of the ciphertext and the two encryption keys [16]. The general scheme of this attack involved two stages: in the first stage, iterative phase retrieval was used to obtain the approximate intermediate function obtained after the first Fourier transformation. In the second stage, the approximate intermediate function and the first encryption key were used to recover the plaintext using a second cycle of iterative phase retrieval algorithm. A similar iterative Fourier transform-based phase retrieval has been shown to breach a double-image encryption system based on the phase truncation method [17].

Different specific attack schemes have been proposed considering different combinations of information known to the attacker. It has been shown that the specific attack on the phase truncation-based system can be carried out using the information of only one of the encryption keys, i.e. the second key [18, 19]. In such an attack, the first stage involves a phase retrieval algorithm based on the HIO algorithm combined with the ER algorithm. The support is used as the constraint to update the approximations in each step. The intermediate function is obtained as this phase retrieval process converges. This intermediate function and the first encryption key are used to retrieve the plaintext.

The attack algorithm begins by initiating the ERA cycle, wherein the ciphertext is multiplied with a random phase. For the kth iteration, this step can be written as

$$E_k(x, y) = E_0(x, y) \times e^{i\varphi_k(x, y)} \tag{11.15}$$

The obtained complex amplitude is subjected to the Fourier transformation, which gives the guess solution for the intermediate function

$$g_k'(u, v) = \mathfrak{F}\{E_k(x, y)\} \tag{11.16}$$

In any ERA cycle, the constraints applied on the object domain are that the object support pixels are retained while the rest of them are set to zero.

$$g_{k+1}(u, v) = g_k'(u, v) \times S \tag{11.17}$$

Here, S represents the object support. The modified solution is inverse Fourier transformed to reach the ciphertext domain.

$$G_{k+1}'(x, y) = \mathfrak{F}^{-1}\{g_{k+1}(u, v)\} \tag{11.18}$$

The amplitude of $G_{k+1}'(x, y)$ is substituted with the ciphertext, while its phase is retained. This gives the input to the next $(k+1)$th iteration.

$$E_{k+1}(x, y) = E_0(x, y)e^{i\varphi_{k+1}(x, y)} \tag{11.19}$$

Here,

$$\varphi_{k+1} = \text{phase}\left[G_{k+1}'(x, y) \right] \tag{11.20}$$

One cycle of the above ERA is followed by 39 cycles of the HIO algorithm. The input to the HIO algorithm is the entity obtained in equation (11.19). This obtained entity $E(x,y)$ is Fourier transformed in the first step of the HIO algorithm. For any nth iteration, this can be written as

$$h_n'(u, v) = \mathfrak{F}\{E(x, y)\} \tag{11.21}$$

In the object domain, the constraints are applied. The pixels following the object support are retained while those outside the object support are provided with a negative feedback to the guess solution obtained in a previous step.

$$h_{n+1}(u, v) = \begin{cases} h_n'(u, v) & \text{otherwise} \\ h_n(u, v) - \beta h_n'(u, v) & (u, v) \in \gamma \end{cases} \tag{11.22}$$

Here, γ represents all those points not following the object constraints. β is a positive-valued parameter whose value lies in the range $(0,1)$. This modified solution is again inverse Fourier transformed to reach the ciphertext domain,

$$H'_{n+1}(x, y) = \mathfrak{F}^{-1}\{h_{n+1}(u, v)\} \tag{11.23}$$

The amplitude is substituted with ciphertext while the phase is retained.

$$E_{n+1}(x, y) = E_0(x, y)e^{i\theta_{n+1}(x, y)} \tag{11.24}$$

$$\theta_{n+1} = \text{phase}\{H'_{n+1}(x, y)\} \tag{11.25}$$

After completion of 39 cycles of the HIO algorithm, the algorithm again enters the ER algorithm. This $(1 + 39)$ cycles constitute a single iteration in context of the entire algorithm. At the end of each iteration a cost function is evaluated, namely the sum squared error (SSE). As the *SSE* converges to a predefined minimum, the solution to the iterative algorithm converges. The SSE in this case is taken to be the error between the actual ciphertext and the approximate ciphertext obtained at the end of each iteration.

$$SSE = 10 \log \frac{\sum_{i=1}^{M}\sum_{j=1}^{N}|H'(x, y) - E_0(x, y)|^2}{M \times N} \tag{11.26}$$

A total of 3000 of such iterations were carried out to get the converged solution, i.e. the intermediate function. Once the intermediate function was obtained, another cycle of phase retrieval can be set up using the knowledge of the first encryption key and the intermediate function to obtain the plaintext image.

The simulation results for specific attack applied on the PTFT-based asymmetric cryptosystem using the information of single encryption key has been shown in figure 11.1 [16]. Figure 11.1(a) shows the binary image used as the plaintext. Figures 11.1(b) and (c) show the RPMs used as K_1 and K_2. The constraint used in the algorithm is shown in figure 11.1(d). The ciphertext is shown in figure 11.1(e) and the retrieved image using the specific attack is shown in figure 11.1(f). Figure 11.1(g) shows the SSE evaluated as a function of the iteration number. It can be seen that it converges to a minimum value after 2500 iterations.

Another major asymmetric optical cryptosystem to be introduced is the EMD-based encryption. Though this method was proposed to counter the vulnerability of the PTFT-based asymmetric cryptosystems to the specific attack, yet it was later shown that the EMD class of cryptosystems too were prone to a certain type of specific attack. In the EMD-based encryption system, the plaintext image is multiplied with an RPM and then subjected to Fourier transformation. The Fourier spectrum is then split into two complex components, with equal amplitudes, one which serves as the ciphertext while the second serves as the key. This is shown in figure 11.2, where the variable Z denotes the Fourier spectrum in the Argand plane. It is split into two components CT (which is used as the ciphertext), and PK

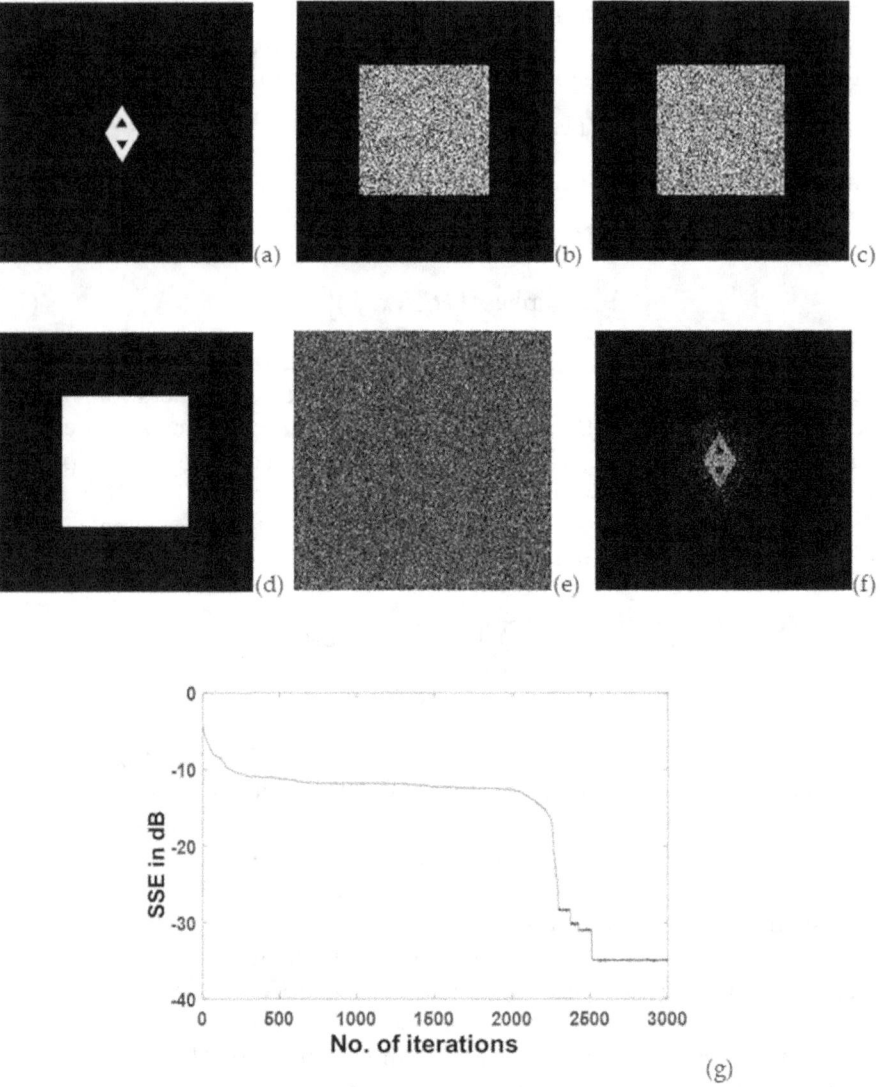

Figure 11.1. Simulation results for the specific attack applied on the PTFT-based asymmetric cryptosystem using the information of single encryption key. (a) Binary image used as a plaintext, (b) intensity image of RPM, K_1, (c) intensity image of RPM, K_2, (d) constraint used in the algorithm, (e) intensity image of ciphertext, (f) retrieved image using the specific attack, and (g) SSE evaluated as a function of the iteration number.

(which serves as the private key). These two entities are represented mathematically as

$$\mathrm{CT} = \frac{Z/2}{\cos(\varphi - \delta)} \exp(i\delta) \qquad (11.27)$$

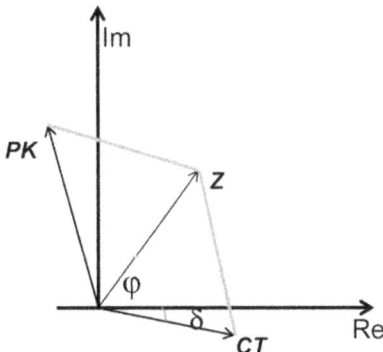

Figure 11.2. EMD represented in the Argand diagram. PK: private key, CT: ciphertext.

$$PK = \frac{Z/2}{\cos(\varphi - \delta)} \exp(i(2\varphi - \delta)) \qquad (11.28)$$

The angle φ represents the phase part of the Fourier spectrum Z i.e., $\varphi = \arg(Z)$. The angle δ is taken randomly as $\delta = 2\pi$ r and (m), r and (m) being a random distribution between [0,1].

It can be seen from equations (11.27) and (11.28) that the private key is interdependent. The fact that ciphertext and the private key have equal amplitudes, along with the knowledge of the RPM (encryption key) is enough for any attacker to set up a specific attack using the phase retrieval algorithms. Essentially, the vulnerability of these systems arises because the private keys are not independent of the ciphertext [20, 21]. It has been shown that the angle of decomposition δ used to split the Fourier spectrum into two components can also be used as a constraint in the iterative phase retrieval-based attack algorithm to get a converged solution [22].

11.8 Collision attack

Collision attack in cryptanalysis refers to the situation when two distinct inputs to the cryptosystem yield the same output. Resistance to this attack is desirable to most of the cryptosystems, especially watermarking and authentication-based systems. It has been shown that phase retrieval algorithms can be used to study the vulnerability to collisions from the ciphertext of the watermarked host image [23]. In this attack, the ciphertext, $E(x,y)$ (generated from the DRPE, equation (11.4)) using phase keys $\exp[i2\pi R_1(x, y)]$ and $\exp[i2\pi R_2(u, v)]$ are known to the attacker. It is aimed to generate a set of different phase keys, say $\exp[i2\pi R'_1(x, y)]$ and $\exp[i2\pi R'_2(u, v)]$ and a different input image, say $f'(x, y)$ that gives the same ciphertext $E(x, y)$. Using equation (11.5), this can be stated as. find a real image, $f'(x, y)$ such that the following equation is satisfied,

$$G(u, v) = |G(u, v)| \exp[i\beta(u, v)]$$
$$- \{\Im\{f'(x, y)\exp[i2\pi R'_1(x, y)]\}\}\exp[i2\pi R'_2(u, v)] \qquad (11.29)$$

This can be solved using the GS algorithm, because the amplitude in the two domains is known. Hence, if the ciphertext is known then any attacker can use this collision attack to produce different ciphertexts that give different plaintexts, which can cause a security breach. Collision attack has been shown to be applicable to the asymmetric cryptosystem based on a phase-truncated Fresnel transform [24].

11.9 Occlusion attack

Occlusion refers to obstructing some of the information in the given ciphertext. The cryptosystem is said to be resistant to the occlusion attack if even after partial information loss of the ciphertext, the plaintext can be retrieved with less noise. In many of the reported cryptosystems, a certain portion of the ciphertext is rendered zero to occlude the information [25, 26]. The ciphertext is then subjected to decryption using the correct keys. A robust cryptosystem yields sufficient information of the plaintext, and is said to resist the occlusion attack.

11.10 Effects of additive and multiplicative noise

Ciphertexts can be contaminated during the communication/transmission process. For an encryption technique to be strong it is desirable that even after such contamination, it is possible to retrieve sufficient information after decryption. Therefore, cryptosystems are checked for their strength by adding some noise to the ciphertext and then analysing the decrypted image with correct keys. In a basic approach to mimic this contamination, additive Gaussian noise and multiplicative or speckle noise are added to the ciphertext and then decryption was carried out [25, 27]. Different parameters, such as the CC value and the MSE are then evaluated to establish the semblance to the plain image. Similarly, other noise such as the Poisson noise can be used to study the strength of the cryptosystem.

References

[1] www.crypto-it.net/eng/theory/kerckhoffs.html
[2] Frauel Y, Castro A, Naughton T J and Javidi B 2007 Resistance of the double random phase encryption against various attacks *Opt. Express* **15** 10253–65
[3] Guo C, Muniraj I and Sheridan J T 2016 Phase retrieval-based attacks on linear canonical transform-based DRPE systems *Appl. Opt.* **55** 4720–8
[4] Toughi S, Fathi M H and Sekhavat Y A 2017 An image encryption scheme based on elliptic curve pseudo random and advanced encryption system *Sig. Process* **141** 217–27
[5] Peng X, Chang P, Wei H and Yu B 2006 Known plaintext attack on optical encryption based on double random phase keys *Opt. Lett.* **31** 1044–6
[6] Gopinathan U, Monaghan D S, Naughton T J and Sheridan J T 2006 A known-plaintext heuristic attack on the Fourier plane encryption algorithm *Opt. Express* **14** 3181–6
[7] Liu W, Yang G and Xie H 2009 A hybrid heuristic algorithm to improve known-plaintext attack on Fourier plane encryption *Opt. Express* **17** 13928–38
[8] Barrera J F, Vargas C, Tebaldi M, Torroba R and Bologini N 2010 Known plaintext attack on a joint transform correlator encrypting system *Opt. Lett.* **35** 3553–5

[9] Tashima H, Takeda M, Suzuki H, Obi T, Yamaguchi M and Ohyama N 2010 Known plaintext attack on double random phase encoding using fingerprint as key and a method for avoiding the attack *Opt. Express* **18** 13772–81

[10] Rajput S K and Nishchal N K 2013 Known-plaintext attack on encryption domain independent optical asymmetric cryptosystem *Opt. Commun.* **309** 231–5

[11] Peng X, Wei H and Zhang P 2006 Chosen-plaintext attack on lensless double random phase encoding in Fresnel domain *Opt. Lett.* **31** 3261–3

[12] Barrera J F, Vargas C, Tebaldi M and Torroba R 2010 Chosen plaintext attack on a joint transform correlator encrypting system *Opt. Commun.* **283** 3917–21

[13] Zhang Y, Xiao D, Wen W and Liu H 2013 Vulnerability to chosen-plaintext attack of a general optical encryption model with the architecture of scrambling-then-double random phase encoding *Opt. Lett.* **38** 4506–9

[14] Carnicer A, Usategui M M, Arcos S and Juvells I 2005 Vulnerability to chosen-cyphertext attacks of the optical encryption schemes based on double random phase keys *Opt. Lett.* **30** 1644–6

[15] Kumar P, Kumar A, Joseph A and Singh K 2009 Impulse attack free double random phase encryption scheme with randomized lens-phase functions *Opt. Lett.* **34** 331–3

[16] Wang X and Zhao D 2012 A special attack on the asymmetric cryptosystem based on phase-truncated Fourier transforms *Opt. Commun.* **285** 1078–81

[17] Fatima A and Nishchal N K 2016 Discussion on comparative analysis and a new attack on optical asymmetric cryptosystem *J. Opt. Soc. Am.* A **33** 2034–40

[18] Nishchal N K and Fatima A 2018 Phase retrieval in optical cryptography chapter 3 *Cryptographic and Information Security* ed S Ramakrishnan (Boca Raton, FL: CRC Press)

[19] Ding X, Yang G and He D 2015 A simple public-key attack on phase-truncation-based double-images encryption system *Opt. Commun.* **346** 141–8

[20] Wu J, Liu W, Liu Z and Liu S 2015 Cryptanalysis of an 'asymmetric optical cryptosystem based on coherent superposition and equal modulus decomposition *Appl. Opt.* **54** 8921–4

[21] Fatima A and Nishchal N K 2018 Equal modulus decomposition based asymmetric optical cryptosystems chapter 5 *Advanced Secure Image Processing for Communications* ed A AlFalou (Bristol: IOP Publishing)

[22] Wang Y, Quan C and Tay C J 2016 New method of attack and security enhancement on an asymmetric cryptosystem based on equal modulus decomposition *Appl. Opt.* **55** 679–86

[23] Situ G, Monaghan D S, Naughton T J, Sheridan J T, Pedrini G and Osten W 2008 Collision in double random phase encoding *Opt. Commun.* **281** 5122–5

[24] Mehra I, Rajput S K and Nishchal N K 2013 Collision in Fresnel domain asymmetric cryptosystem using phase truncation and authentication verification *Opt. Eng.* **52** 028202

[25] Nishchal N K, Joseph J and Singh K 2004 Fully phase-based encryption using fractional order Fourier domain random phase encoding: error analysis *Opt. Eng.* **43** 2266–73

[26] Fatima A, Mehra I and Nishchal N K 2016 Optical image encryption using equal modulus decomposition and multiple diffractive imaging *J. Opt.* **18** 085701

[27] Rajput S K and Nishchal N K 2012 Image encryption and authentication verification scheme using fractional nonconventional joint transform correlator *Opt. Lasers Eng.* **50** 1474–83

IOP Publishing

Optical Cryptosystems

Naveen K Nishchal

Chapter 12

Optical security keys/masks

12.1 Introduction

The requirements of a security system are: (a) significant length of the key and its randomness and (b) exact regeneration during decryption. Therefore, in any security system, keys play a significant role. The security systems are designed in such a way that only with the use of proper keys the original information should be retrieved. Hence, the success of any such system fully depends on the key design and its implementation. In other words, the security of an encryption system depends on how secure and strong, the key is. As discussed in previous chapters, it is always desired to have a larger key size in an encryption system for achieving a higher level of security. Optical encryption techniques differ with digital counterparts in terms of key design and algorithm. In optical systems, mostly a physical key is used. The various encoding schemes include the use of random phase, polarization sensitive techniques and different types of digital methods, etc. In optical cryptosystems, the keys are mostly phase-only functions called RPMs [1]. Speckle patterns are the classical examples of RPMs. The phase masks cannot be read, photographed, or copied as easily as intensity masks. As discussed in chapter 2, the RPMs used in a cryptosystem are statistically independent to each other. It is for this reason it is believed that optical cryptosystems provide a more complex environment and are more resistant to attacks than are purely digital systems.

In the DRPE technique, security of the key is related to the space-bandwidth product of the used RPMs. The space-bandwidth product of the RPMs can be increased only to the extent that it does not exceed the space-bandwidth product offered by the optical system. RPMs are characterized by their auto-correlation and cross-correlation properties. Depending upon the design, there are various types of amplitude and phase-only masks used in optical cryptosystems.

- Diffractive optical element (DOE)
- Holographic mask (HD)
- Random amplitude mask (RAM)

- Random phase mask (RPM)
- Structured phase mask (SPM)
- Plasmonic phase mask (PPM)
- Plastic diffuser (PD)
- Ground glass diffuser.

The creation of encryption keys with high security is crucial to any cryptosystem. Therefore, generation of keys is a challenging topic of research. This chapter discusses the various key designs, synthesis, and their features.

12.2 Literature review

Numerically generated RPMs have been proposed in the conventional DRPE system [1]. Two statistically independent RPMs, $\exp\{i2\pi R_1(x,y)\}$ and $\exp\{i2\pi R_2(u,v)\}$, are employed at the input and frequency plane, respectively. An input image/data is transformed into a complex-valued Gaussian stationary white noise that provides maximum entropy, which is called the encrypted image/data. $R_1(x,y)$ and $R_2(u,v)$ are two independent random numbers uniformly distributed in [0,1]. The properties of RPMs have been discussed in chapter 2. Since DOE designing has commonality with the image recovery problems, employing a modified GS algorithm, two cascaded diffractive structures were designed to be used in optical cryptosystems [2]. Single or multiple cascaded phase arrays have been designed to produce a desired intensity distribution at a given transverse plane.

In order to preclude the counterfeiting of document data, optical memories have been used, in which the data is represented by an invisible optical code [3]. A 2D ultra-violet bar code is a simple example of optical memory, which is used in document security. Due to the promises of optical systems based on holographic memory, diffractive optical memories have received increased attention. Such memories are useful in high storage capacity, high data transfer rate, and fast data seek time. This technology finds applications in coding, reading, checking, assessing authenticity, and development and recognition of cost efficient and secure authenticity marks. The diffractive structures are resilient against reorigination and can be physically destroyed using irreversible writing techniques. Therefore, they can be easily incorporated into documents using glues.

A secured optical memory using the DRPE technique has been demonstrated [4]. In this study, the RPMs were prepared by bleaching gray-scale photographic films of uniformly distributed white noise. The phase value of each pixel was made independent of the other pixels distributed on $[0, 2\pi]$. A 2D gray-scale matrix of a random array uniformly distributed in [0, 255] was displayed on a monitor, which was then photographed. In order to make sure that the phase value lies within $[0, 2\pi]$, a three-level scale of 0, 127, and 255 gray-values were displayed along the side of the random array. After bleaching, the phase values $(0, \pi, \text{ and } 2\pi)$ corresponding to the scaled gray-values (0, 127, and 255) were measured by interferometry. By adjusting the conditions of the RPM making process, the desired phase values were achieved.

Electron-beam lithography techniques have been used in fabricating diffractive optical elements (DOEs) and CGHs for security and authentication verification applications. CGH designed with multiple image planes from the image region has been reported [5]. There are two types of binary CGH coding method: cell oriented and point oriented. In the cell-oriented method a hologram is divided into many cells depending upon the sampling points. On the other hand, in the point-oriented method, a sampling point is represented by a single aperture. The cell-oriented method is cost effective as compared to the point-oriented method.

An optical cryptosystem using phase conjugation in a photorefractive crystal has been reported, in which the effect of finite space-bandwidth product of RPM on the encryption system's performance was also elaborated [6]. In this study, the RPMs were prepared by depositing gelatin nonuniformly upon a glass plate such that its thickness across the plates varies in a random manner [7]. Practically, it is possible to use an SLM in the frequency plane for displaying RPM, where the RPM can be considered to be constituted of regions called correlation cells in which the phase value remains the same [8]. In an SLM, the minimum size of a correlation cell is represented by the pixel size of the SLM. The space-bandwidth product of an RPM is defined as the ratio of the area of the mask to the area of the mask cell. An SLM with $N \times M$ pixels and each pixel having finite width can take [0, 255] gray values. Detailed mathematical analysis has been carried out considering three cases when RPMs used during encryption were used in decryption or there was some shift introduced into the RPM during decryption [6].

A phase-only encryption under DRPE framework has been demonstrated, in which a binary phase key was implemented electronically on a phase-only SLM [9]. A 2D array of random phase values generated electronically scrambled the input information. An image encryption using phase-shifting interferometry has been demonstrated, in which a commercially available plastic diffuser of randomly varying thickness was used as the RPM. The plastic diffuser had the correlation length of 6 μm in both the x- and y-directions [10]. An image decryption system has been reported that used interference of a reference wave passing through the encrypted image and a phase-only decryption key, made from a linear phase summation of 2π with the RPM applied during encryption [11]. The phase-only decrypting key was fabricated by optical lithography. A fully-phase encryption in the FRT domain under the DRPE framework has also been reported, in which RPMs were prepared by removing, using a fixer, the silver halide emulsion of a holographic plate [12]. In other studies, RPMs have been prepared by exposing a holographic plate with a fully developed speckle pattern obtained from a diffuser. The holographic plate is illuminated with a coherent beam of light. The exposed plates were processed (developed and fixed) and bleached [13, 14].

An image encryption method under the DRPE framework using zone plate as RPM has been reported [15]. The phase mask used in the study was actually a toroidal zone plate, called a structured phase mask (SPM), which is basically a DOE. A fractal zone plate has also been used as an RPM [16]. The SPMs have a shape given by a specific configuration. Such phase masks have some benefit as they can be easily positioned during the decoding process, their diffraction efficiency can be manipulated depending upon the requirement of the optical system, and the generation involves multiple parameters that serve as multiple keys in a security system. A toroidal zone plate has its own centering mark (on the axis focal ring when illuminated by a parallel beam), which makes the alignment problem easier in actual implementation [15–18].

Multiple image encryption architecture employing different pupil aperture masks in the encoding lens has been reported [19]. The geometrical parameters characterizing the apertures served as the additional security keys. Considering N different pupil aperture masks in the encoding lens, N different input images can be individually processed, one for each aperture. In this study, the size of the aperture mask was 15 mm × 11 mm. The used RPM was a commercially available standard pure phase glass diffuser. The encryption architecture was implemented in a phase conjugation geometry, in which it was shown that with the decrease in pupil aperture size the phase conjugation reflectivity increases. Also, by modifying the pupil aperture diameter, the speckle size can be adjusted. In a subjective speckle, the average speckle size is given as [19]

$$\langle S_z \rangle \approx \lambda \left(\frac{d_c}{D} \right)^2 \tag{12.1}$$

where d_c is the distance between the imaging lens and the photorefractive crystal plane, λ denotes the operating wavelength, and D stands for the diameter of the aperture lens.

An encryption procedure employing a sandwich diffuser in the Fourier plane has also been reported, in which a diffuser was prepared after sandwiching two RPMs together [20]. In this study, it was shown that the decryption was not possible with the use of only one of the two RPMs constituting the sandwich diffuser. Also, if any of the diffusers constituting the sandwich was shifted in-plane with respect to the other, no decryption is achieved. Due to double diffraction, a single sandwich diffuser offers a larger key space in the cryptosystem. But the alignment of two diffusers making a sandwich diffuser is practically a difficult issue. Two identical phase diffusers were prepared by exposing two holographic plates with the same speckle pattern. These two diffusers were sandwiched together and illuminated with a parallel beam of light. Applying in-plane shift to one of the phase diffusers of a sandwich diffuser, Young's fringes were obtained in the Fourier plane, as shown in figure 12.1.

Emphasizing the miniaturization, a nanoworld-based security system has been reported [21]. The key generation involved a physical nano-object and its characterization process. It was shown that using nano-objects and optical response, a complex optical tomography map can be produced, which could be employed as a pseudorandom generator. The secret key was generated through the near-field optical data of nano-objects. It was stated that the nano-based secret key must be destroyed and regenerated from the nano-object at request, when decryption is required. This will reduce the risk of hacking. A plasmonic key-based encryption scheme has also been reported [22, 23]. The sensitivity to the initial conditions of the

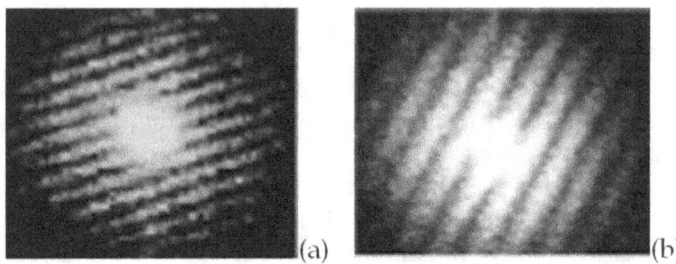

Figure 12.1. Fringe patterns generated by a sandwich diffuser for (a) 60 μm in-plane shift and (b) 40 μm in-plane shift in the x–y plane of one of the diffusers of a sandwich diffuser. Reproduced with permission from [20].

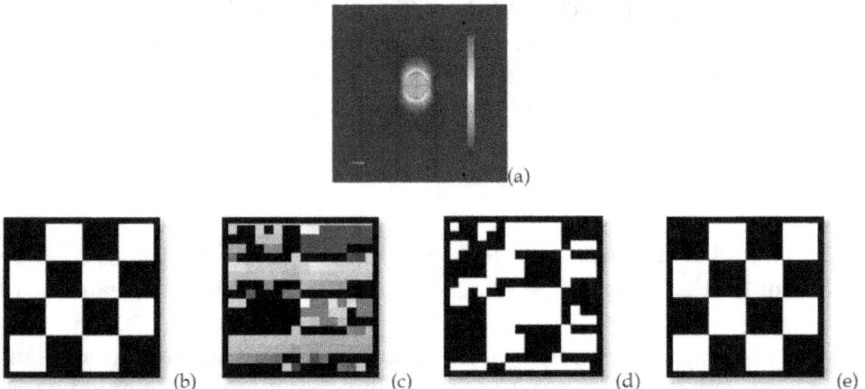

Figure 12.2. Simulation results for XOR encryption; (a) plot of the electric field norm near the nanosphere surface of radius 10 nm, (b) input image of size 16×16 pixels, (c) plasmonic mask constructed with $\lambda_1 = 525$ nm, $r_1 = 10$ nm and $\lambda_2 = 500$ nm, $r_2 = 15$ nm, (d) the encrypted image using XOR encoding scheme, and (e) the decrypted image obtained after using the correct keys.

plasmonic mask introduces physical parameters such as the radius of the sphere, the wavelength of the impinging beam, and the shape of the nanostructure, as the key. The proposed plasmonic system consisted of a nanosphere illuminated with an electromagnetic beam polarized in the z-direction. The optical map generated from the simulation of scattering of the electromagnetic wave from the nanosphere formed the prototype of the plasmonic mask. The electric field values obtained from various spatial points were used to form the plasmonic mask, which was used in the logical XOR encryption [23]. The results of numerical modeling of the nanosphere carried out using COMSOL software was based on the finite element method. The simulation results carried on the Matlab platform have been shown in figure 12.2.

Multiple-image encryption using the nonlinear iterative phase retrieval algorithm in the GT domain has been reported, in which an optical vortex beam was used for illumination. Two POMs were generated, for each input image, using two chaotic SPMs based on a logistic map, having larger randomness [24]. It was claimed that

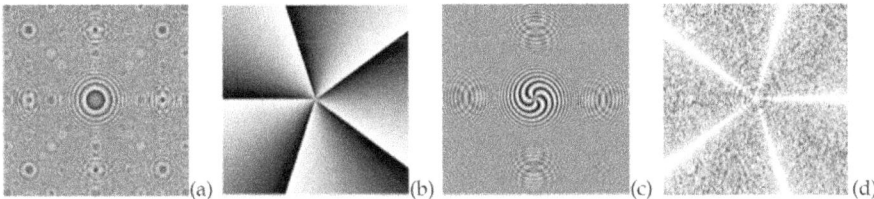

Figure 12.3. Simulation results for different encryption keys. Generated SPM based on (a) Fresnel zone plate, (b) radial Hilbert mask, (c) Fresnel zone plate and radial Hilbert mask, and (d) chaotic SPM. Reprinted with permission from [24]. © The Optical Society.

the use of chaotic SPMs help ease the alignment problem. Figure 12.3 presents the numerically generated different SPMs.

An encryption scheme based on a deterministic mask, which is again a class of SPM, has been reported [25]. The deterministic phase masks were built from a linear combination of several subkeys. A cryptosystem with incoherent ultra-broadband illumination has also been reported that used a commercially available ground glass diffuser (Edmund, 120 Grit Ground Glass Diffuser) as an RPM [26]. Following a different strategy, an encryption system using RPMs based on gamma distribution in the gyrator domain has been reported [27]. Using random parameters according to the gamma probability distribution function, two random distribution phase images were generated. The different parameters of gamma distribution RPMs modulated the statistical distribution of phase in the ciphertext. Recently, liquid surface patterns have been used as RPMs in an optical encryption system [28]. It has been shown that the encrypted images maintain maximum entropy. Also, the system has the capability to resist the KPA due to the time-dependence from the liquid surface patterns. Using a liquid system (a thin liquid system with a controllable surface topology) DRPE can have fixed and static phase masks. Thus, the system can have a time-dependent transfer function according to the user-defined liquid surface. The applicable attacks, which depend on plaintext-ciphertext pairs, can be easily avoided in this case.

Broadly speaking, phase keys used in optical cryptosystems can be categorized into two types: non-structured phase masks and structured phase masks. The features of an SPM can be controlled through various parameters during fabrication. An RPM is an example of a non-structured phase mask, while SPMs are basically DOEs or holographic optical elements (HOEs). DOEs are produced either as amplitude-only DOEs or phase-only DOEs. The phase-only DOEs can be produced using several phase levels. Another type of phase mask has also been reported which is analytically generated through an iterative phase retrieval algorithm. Such phase masks are mostly object-dependent.

12.3 Random phase mask

RPMs are important technical elements used in various photonic security applications. In the 1970s, the RPMs were used in recording the Fourier transform holograms [29, 30]. In the Fourier transform plane, most of the light is concentrated

Figure 12.4. Intensity image of a numerically generated (on the Matlab platform) random phase code.

in an array of bright spikes and the use of RPM helps spread the light distribution over the entire Fourier transform plane. The intensity of the output light beam is not affected by the RPM because the phase mask only changes the phase not the amplitude of the transmitted light beam. This idea was borrowed into making optical cryptosystems. An RPM is mathematically expressed as $\exp\{i2\pi R(x,y)\}$, in which $R(x,y)$ is a white sequence uniformly distributed in [0,1]. In an encryption system, usually more than one RPM is used. All the used RPMs should be statistically independent from each other. Speckle patterns are the classical examples of RPMs.

Different methods of fabricating RPMs have been reported in literature. Nowadays, commercially available ground glass diffusers or plastic diffusers are also used. Displaying numerically generated random phase values onto a phase-only SLM is a convenient technique. The larger size of the RPM offers a higher level of security. It has been recommended that in an optical cryptosystem, the minimum size of an RPM should be 1024×1024 pixels and a single RPM should not be repeatedly used for securing several images/data. Figure 12.4 shows an intensity image of a numerically generated (on Matlab platform) random phase code. For generating random numbers in Matlab, *rand* or *randn* functions are used. The function *rand* generates uniformly distributed random numbers in the interval [0,1] while *randn* generates normally distributed random numbers.

12.4 Structured phase mask

SPMs based on the Fresnel zone plate, fractal zone plate, toroidal zone plate, radial Hilbert mask, spiral phase mask, and deterministic phase mask have been reported [15–18]. All such masks are basically DOEs or HOEs. A toroidal wavefront can be written as [15]

$$U(r) = \exp\left\{\frac{-ik}{2f_T}(r - r_0)^2\right\}$$
(12.2)

where it is assumed that the axis of propagation coincides with the optical axis. $k = 2\pi/\lambda$, where λ stands for operating wavelength, and f_T and r_0 are positive real

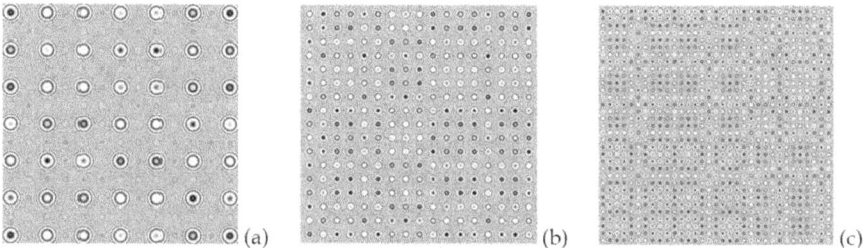

Figure 12.5. Three numerically generated zone plates used as the encryption key. The SPM construction parameters used in the simulation are, for SPM1, f = 150 mm; number of pixels = 1024. (a) SPM1 generated with pixel spacing = 10 μm; λ = 632.8 nm, (b) SPM2 generated with pixel spacing = 15 μm; λ = 532 nm, and (c) SPM3 generated with pixel spacing = 20 μm; λ = 488 nm.

constants. A zone plate has been used as an SPM, which is mathematically expressed as [17],

$$S(x,y) \;=\; t_0 \cos\left(\pi\frac{x^2 + y^2}{\lambda f}\right) \tag{12.3}$$

where, t_0 is a constant, λ is the wavelength of the light source, and f is the focal length of the zone plate. The construction of a zone plate depends on the wavelength of the light source. Wavelength-dependent SPMs have been used in securing a color image [17]. Figure 12.5 shows three numerically generated SPMs constructed for three different wavelengths.

After more than 25 years of development, the family of optical information encryption systems has greatly expanded. In such systems, keys play a vital role. It will not be an exaggeration to state that it is the key of a security system that decides its applicability and reliability. Therefore, optical cryptosystems continue to evolve towards the quicker and safer transformation of information.

References

[1] Refregier P and Javidi B 1995 Optical image encryption based on input plane encoding and Fourier plane random encoding *Opt. Lett.* **20** 767–9

[2] Johnson E G and Brasher J D 1996 Phase encryption of biometrics in diffractive optical elements *Opt. Lett.* **21** 1271–3

[3] Tompkin W R and Staub R 1996 Low-density diffractive optical memories for document security *Opt. Eng.* **35** 2513–8

[4] Javidi B, Zhang G and Li J 1997 Encrypted optical memory using double-random phase encoding *Appl. Opt.* **36** 1054–8

[5] Yoshikawa N, Itoh M and Yatagai T 1998 Binary computer generated holograms for security applications from a synthetic double-exposure method by electron-beam lithography *Opt. Lett.* **23** 1483–5

[6] Unnikrishnan G, Joseph J and Singh K 1998 Optical encryption system that uses phase conjugation in a photorefractive crystal *Appl. Opt.* **37** 8181–5

[7] Unnikrishnan G, Joseph J and Singh K 2001 Fractional Fourier domain encrypted holographic memory by use of an anamorphic optical system *Appl. Opt.* **40** 299–306

[8] Kral E L, Walkup J F and Hagler M O 1982 Correlation properties of random phase diffusers for multiplex holography *Appl. Opt.* **21** 1281–90

[9] Mogensen P C and Glueckstad J 2000 Phase-only optical encryption *Opt. Lett.* **25** 566–8

[10] Tajahuerce E, Matoba O, Verrall S C and Javidi B 2000 Optoelectronic information encryption with phase-shifting interferometry *Appl. Opt.* **39** 2313–20

[11] Seo D-H and Kim S-J 2003 Interferometric phase-only optical encryption system that uses a reference wave *Opt. Lett.* **28** 304–6

[12] Nishchal N K, Joseph J and Singh K 2003 Fully phase encryption using fractional Fourier transform *Opt. Eng.* **42** 1583–8

[13] Nishchal N K, Joseph J and Singh K 2003 Optical phase encryption by phase contrast using electrically addressed spatial light modulator *Opt. Commun.* **217** 117–22

[14] Nishchal N K, Joseph J and Singh K 2004 Securing information using fractional Fourier transform in digital holography *Opt. Commun.* **235** 253–9

[15] Barerra J F, Henao R and Torroba R 2005 Optical encryption method using toroidal zone plate *Opt. Commun.* **248** 35–40

[16] Tebaldi M, Furlan W D, Torroba R and Bolognini N 2009 Optical data storage readout technique based on fractal encrypting masks *Opt. Lett.* **34** 316–8

[17] Rajput S K and Nishchal N K 2012 Asymmetric color cryptosystem that uses polarization selective diffractive optical element and structured phase mask *Appl. Opt.* **51** 5377–86

[18] Singh H, Yadav A K, Vashisth S and Singh K 2015 Double phase image encryption using gyrator transforms and structural phase masks in the frequency plane *Opt. Lasers Eng.* **67** 145–56

[19] Barerra J F, Henao R, Tebaldi M, Torroba R and Bolognini N 2006 Multiple image encryption using an aperture-modulated optical system *Opt. Commun.* **261** 29–33

[20] Singh M and Kumar A 2007 Optical encryption and decryption using sandwich random phase diffuser in the Fourier plane *Opt. Eng.* **46** 055201

[21] Grosges T and Barchiesi D 2010 Toward nanoworld-based secure encryption for enduring data storage *Opt. Lett.* **35** 2421–3

[22] Francois M, Grosges T, Barchiesi D and Erra R 2011 Generation of encryption keys from plasmonics *PIERS Online* **7** 296–300

[23] Fatima A, Mehra I and Nishchal N K 2014 Plasmonics-based keys for optical image encryption *Int. Conf. Fibre Optics and Photonics* OSA Technical Digest S5A.52

[24] Sui L, Bei Z, Xiaojuan N and Ailing T 2016 Optical image encryption based on the chaotic structured phase masks under the illumination of a vortex beam in the gyrator domain *Opt. Express* **24** 499–515

[25] Zamrani W, Ahouzi E, Lizana A, Campos J and Yzuel M J 2016 Optical image encryption technique based on deterministic phase masks *Opt. Eng.* **55** 1031081

[26] Sahoo S K, Tang D and Dang C 2017 Enhancing security of incoherent optical cryptosystem by a simple position-multiplexing technique and ultra-broadband illumination *Sci. Rep.* **7** 17895

[27] Sun W, Wang L, Wang J, Li H and Wu Q 2018 Optical image encryption using gamma distribution phase masks in the gyrator domain *J. Eur. Opt. Soc. Rapid Publications* **14** 28

[28] Schipf D R and Wang W-C 2018 Optical encryption using a liquid phase mask *OSA Continuum* **1** 1026–40

[29] Burckhardt C B 1970 Use of a random phase mask for the recording of Fourier transform holograms of data masks *Appl. Opt.* **9** 695–700

[30] Stewart W C, Firester A H and Fox E C 1972 Random phase data masks: fabrication tolerance and advantages of four phase level masks *Appl. Opt.* **11** 604–8

www.ingramcontent.com/pod-product-compliance
Lightning Source LLC
Chambersburg PA
CBHW080912170526
45158CB00008B/2082